Game for Anything

Game for Anything
Writings on Cricket

Gideon Haigh

Published by Black Inc.
An imprint of Schwartz Publishing

Level 5, 289 Flinders Lane
Melbourne Victoria 3000 Australia
email: enquiries@blackincbooks.com
http://www.blackincbooks.com

National Library of Australia Cataloguing-in-Publication entry:

Haigh, Gideon.
Game for anything : writings on cricket.

ISBN 1 86395 309 4.

1. Cricket - Anecdotes. 2. Cricket players - Australia.
I. Title.

595.726

Book design: Thomas Deverall

Printed in Australia by Griffin Press

Contents

Introduction

WHEN THE AMERICAN SHORT STORY WRITER Ring Lardner died, young, after a career that never quite fulfilled its rich promise, F. Scott Fitzgerald was quick to finger the culprit. Lardner, he pointed out, had poured much of his youthful energy into working as a beat reporter for the *Chicago Inter-Ocean* following the Cubs and White Sox, and thus down the drain of sports journalism: 'During those years when most men and women of promise achieve an adult education, Ring moved in the company of a few dozen illiterates playing a boy's game. A boy's game, with no more possibilities in it than a boy could master, a game bounded by walls which kept out novelty or danger, change or adventure.' So there.

In the years I have written about cricket, Fitzgerald's admonition has always had a faintly familiar ring. Why bother doing it? Would there not be more gainful avenues for expression? At book launches and literary events, a hint of condescension lurks behind the smiles. Still writing about cricket, eh? *Oh dear.* There is even, at times, a hint of resentment, sport in Australia being seen as somehow antagonistic to culture, an alternative to thought, an opiate of the people. A curious country, this – where it is often those boasting loudest of their demotic sympathies who are most inclined to dismiss the widely enjoyed as base, vulgar and jejune.

Which is not to say that these questions are without force. Yes indeed: why do it? When I ask any of the scores of young people I've met aspiring to careers writing about sport for newspapers, the answer is always invariable: 'Because I love sport.' I cringe inwardly at this; perhaps the only answer worse is: 'Because I love writing.' It isn't that enjoyment always ceases once one turns a pleasure into a job; but one's relationship to that pleasure changes. When I meet them again down the track, they have usually become either bored or boring, surfeited and cynical or capable of discussing nothing else – chewed up and spat out or chewed up and swallowed.

Looking back, even if I was incapable of articulating it, I think I've always had in mind the possibility of such a fate. It was love at first sight when I found cricket. I've tried not to jeopardise the relationship. I still play it, humbly, and enjoy it, greatly. Thus I've shrunk from making cricket my job: where I've written, it's usually been because I've had something to say, or at least to add, wished to share it, and thought it time well spent. It's meant to be fun, isn't it? We're doing this because we want to, right? I've always liked a story about a conversation between Gubby Allen, who between captaining England and being a grandee of Lord's worked in the City, and his Middlesex colleague John Warr. 'If I had devoted as much time to stockbroking as I had to English cricket,' Allen declared loftily one day, 'I should probably have been a millionaire by now.' Warr hastily assented. 'Yes,' he said. 'And just think how much better off English cricket would have been.'

This book collects some of my cricket journalism from the last five years – the stuff, anyway, that it's not too shaming to revisit, and hasn't been rendered redundant by that terror of the sportswriter, glorious uncertainty – plus a couple of earlier pieces. A few have been published in Australia before; most have not, being commissioned from England, India and Pakistan. There are studies of players old and new, essays exploring past events and present discontents, satisfactions of my own curiosity and responses to events of the day. Some subjects I have tackled several times, such as Sir Donald Bradman's apotheosis and Hansie Cronje's disgrace. One section addresses lesser cricketers who, for various reasons, have become personal favourites, from 'Ranji' Hordern to Mohinder Amarnath. Another concerns writing and broadcasting, including perspectives on five cricket critics I admire and one I do not. The Yarras, my club, lurk in the background. A few of these pieces made their first appearance in the club newsletter of which I'm editor – still my longest permanent cricket-writing role. But to quote Ring Lardner himself, at the start of his wonderful short story of sport and journalism 'Harmony': 'Even a baseball writer must sometimes work.'

Gideon Haigh
Melbourne, August 2004

ACKNOWLEDGEMENTS

Behind every piece in this book has stood an editor, sometimes pleased, sometimes puzzled, but usually good-humoured: Stephen Fay, Sambit Bal, Chris Ryan, John Stern, Graem Sims, Graeme Wright, Peter Hanlon, Warwick Franks, Rob Steen, Guy Rundle, Peter Rose, Max Suich, Sally Heath, Kathy Bail and Garrie Hutchinson. Special thanks are due to Malcolm Schmidtke, who eased my first faltering steps in journalism, and David Frith, who's forgotten more about cricket than I'll ever know. I benefited, as always, from Philippa Hawker's critical clarity and George Thomas's eagle eye when they read the manuscript. It's been another enjoyable and trouble-free experience to publish with Morry Schwartz and Black Inc: that is, Rebecca Arnold, Eugenie Baulch, Thomas Deverall, Chris Feik, Roisin FitzGerald, Meredith Kelly, Sophy Williams and Caitlin Yates.

The confidence with which many of the judgements in this book have been made testifies to how easy cricket is when perched behind a laptop. I am grateful to my comrades at the Yarras for tolerating my efforts to bridge the gulf between talking and doing, and to my partner Sally Warhaft for dealing with the frustrations that flow from my repeated failures to do so. In all areas other than sport, she has hugely enriched my life. I dedicate this book to her, in lieu – temporarily, I promise – of a summer holiday.

Past Masters

A Brief History of Cricket

FOR ALL THE BOOKS that have purported to document the history of cricket, there is much about its past we do not know, and probably never will.

The bat has been explained as a development of a shepherd's crook for not much more reason than it looks like a shepherd's crook; of the choice of leather, cork and wool to make the ball, we can say little more than that these were the available materials. Two stumps are first mentioned in 1680, and a third appears in a picture a century later, even though it wasn't officially necessary until 1838. How was it decided we bowl with a rigid arm, and to play in opposite directions in alternate overs? Brilliant ideas, both of them, the former encouraging innovation to challenge the rigidity of the requirement, the latter introducing chance and variety by granting environmental degradation an influence on a match's unfolding – but who to credit with them?

Above all, why did cricket catch on? The world abounds in games that don't, or do then die: cricket's popularity in places and with people so far from its origins is a source of ineradicable wonder. For sure, cricket can look spectacular, yet it's not geared to spectacle; it can be entertaining, but it's not really an entertainment. The main action takes place further from watchers than in any other sport, while the density and sophistication of its laws and techniques make few concessions to the uninitiated. Were you to design a game with modern mores in mind, in fact, the last thing you'd come up with would be convoluted, opaque, prolonged, sometimes ungainly, often slow-moving cricket. And yet, here we are, still playing, and seldom crazier for it.

So what do we actually know? Cricket is fond of its ancient origins. In a semi-serious vein, a passage in Book Six of the *Odyssey* has been advanced as the earliest cricket reference: Nausicaa is glimpsed by the seashore passing a ball to one of her maids who 'dropped it instead into the deep and eddying current'. A beach cricket misfield, surely? Less sentimental interpreters, distinguishing between cricket and a range of possible ancestor pursuits, find it hard to stretch its history beyond 350 years. And only in the eighteenth century did the game become a significant mass diversion, celebrated by devotees and mystifying to visitors, like Frenchman Cesar de Saussure who wrote home after witnessing a match in 1728: 'They go into a large open field, and knock a small ball about with a piece of wood. I will not attempt to describe this game to you, it is too complicated.'

It could be argued that cricket's history, as distinct from its pre-history, dates from 1744, in which were published the earliest extant version of its laws, and occurred the first significant match for which a detailed score remains: Kent versus All-England. The venue for that encounter, London's Artillery Ground, was also that year the first we know of to charge admission: 2d. Cricket, then, was being legislated for, recorded and watched for a fee: preconditions for any game of significance.

Gambling, and wealthy patrons who enjoyed it, abetted cricket's spread. Most early sets of cricket laws contain provisions for the fair settlement of bets; it has even been contended that the publication of the first average tables was motivated by the need for a reliable 'form guide'. Alas, as Jefferson said, the more money, the less virtue: by the early nineteenth century, matches were being routinely fixed, gamblers frequently swindled, and the whole rotten system collapsed under its own weight. William Fennex, a leading cricketer who publicly repented crooked ways, confessed: 'Matches were bought and matches were sold, and gentlemen who meant honestly lost large sums of money, till the rogues beat themselves at last. They overdid it; they spoilt their own trade.' Cricket's slump registers in *Score & Biographies*, which documents as many as many as thirty-three important matches in 1798, but by 1811 only two.

It was not gambling per se that left the deepest imprint on cricket: rather was it the decline of gambling. In celebrating its policing and eventual prohibition, aficionadoes first began placing cricket on a lofty moral pedestal. 'The constant habit of betting will take the honesty out of any man,' argued Rev James Pycroft in *The Cricket Field* (1851). 'It is no small praise of cricket that it occupies the place of less innocent sports. Drinking, gambling, cudgel-playing insensibly disappear before a manly recreation, which draws the labourer from the dark haunts of vice and misery to the open common.' It is to Pycroft's paean of praise that we owe that ever-popular expression for the unethical and mean-minded: 'Not cricket.' And – *pace* Phil Knight and Mark McCormack – there has never been a more successful piece of sports marketing. Even after so many offences against cricket's so-called 'spirit', it remains a point of reference: a source of solace and strength for some, an anachronistic irritation to others.

If we are seeking a dawn of modern cricket, there is much to recommend 1864. Overarm bowling became legal, thus completing bowling's 'march of intellect' from underarm innocence – in the opinion of Sir Donald Bradman 'possibly the greatest single change in the development of the game'. A clean-shaven, fifteen-year-old William Gilbert Grace first notified the game of his precocity with a series of big scores – his batting would revolutionise the way cricket was played, and his personality the way it was organised and promoted. That year, too, came the first edition of *Wisden Cricketers' Almanack,* beginning the trans-substantiation of scores into records. Among a host of matches documented was that undertaken by an England XI against an XVIII from the Australian colony of Victoria: for many, the first evidence of cricket beyond the British Isles.

Tracing cricket's global spread is both very simple and very difficult. It was a colonial game, forming part of the baggage of the British administrative, military and mercantile populations, and no one has significantly improved on the historian Bill Mandle's argument that playing cricket allowed those at the imperial periphery to prove their prowess and pluck to those at the centre. How actively the English proselytised is less clear. History's authorised version sees cricket as being gifted to the empire's subject peoples; history's revised edition has the subject peoples appropriating and adapting it; there was probably a bit of both. Cricket flourished most quickly in racially and culturally homogenous Australia; it took on different accents and attitudes in South Africa, India, the West Indies and New Zealand.

Cricket showed two remarkable qualities in this period – and continues to show them. The first was its capacity to serve both imperial and national ends, to be a means for the payment of homage and for the expression of independence. The second was its technical porosity. Grace, the first giant of batting, playing off front and back foot with equal facility, was the quintessential *rosbif*; Ranjitsinhji, the second, who first opened the leg side as a scoring area, and first harnessed the pace of the ball in the power of the stroke, was perhaps the most exotic cricketer ever born, a self-styled prince from Jamnagar with uncanny eyes and wrists. The trade deficit in cricket skill between England and its dominions closed very quickly in the international game's history. In some areas, England had to do the catching up: exposed to a flint-hearted fast bowling pair in 1921, Australia's Jack Gregory and Ted McDonald, it had to wait until the Bodyline series of 1932-33 to retaliate. In other areas, it never did: no batsman, and very few athletes, have remotely paralleled Donald Bradman in efficiency.

For some decades, cricket was only nominally an international game. Bradman played Tests only in Australia and England, and until 1965, Australia had not ceded a series to anyone but England. After it lost consecutive away series to the West Indies and South Africa, life was never the same. In some respects, it was from that moment that the rise of cricket as a global professional sport became inevitable: that only in England could a cricketer earn a livelihood was an anomaly on borrowed time. The reckoning was delayed by the 1968 relaxation of restrictions on imported foreigners in county cricket, but couldn't be averted.

Kerry Packer was no philanthropist. 'That makes me sound more generous than I am,' said the Australian television magnate. Nor was he a globalist. His motivation was a purely domestic commercial ambition: exclusive broadcasting rights for Australian Tests. But his

World Series Cricket – the first fully professional cricket attraction since the heyday of William Clarke's All-England XI in the 1840s and 1850s – ramified far beyond its original sphere of influence. In Packer's hothouse, innovation flourished. Night cricket, coloured clothing, helmets and drop-in pitches were pioneered, and spread far and wide. Cricketers were awakened to their market value. Administrators began to recognise television and sponsorship as an important revenue source, and television and sponsors to exert a gravitational pull on the game's schedules and conduct.

Above all, Packer popularised limited-overs cricket: the one-day international went forth and multiplied, its numbers growing like a particularly virulent strain of bacteria. None had foreseen the success of the first World Cup, in 1975; after Packer, the World Cup grew in extent and ambition, towing along a host of other tournaments in its wake. The greatest beneficiaries of Packer-style professionalism were the West Indies, suddenly provided with rewards commensurate to their talents: cricket's trends played to their strengths of explosive fast bowling, free-scoring batting and elastic fielding. Those who learned most from Packer's enterprise, however, were Indians, gluttons for glamour, smitten with spectacle. Seldom has a single match had such reverberations as India's upset of the West Indies in the final of the 1983 World Cup. Wise judges had held that the short-form game would not prosper on the sub-continent; after Kapil Dev's coup at Lord's, it arguably displaced Test cricket in popularity and certainly in profitability. In the 1987 and 1996 World Cups, held in India, Pakistan and Sri Lanka, the roars of the crowd competed with the k-ching of the cash register.

When India's cricket board secretary Jagmohan Dalmiya crusaded for the chairmanship of the International Cricket Council in July 1996, using as his election platform the rich pickings of the recent World Cup, it was hard to resist the logic of his appointment. India was the Byzantium to England's Rome, the wealth of the former a contrast to the latter's fading grandeur. But Dalmiya's tenure was overshadowed by scandal, as more money again entailed a virtue shrinkage. The problem was not so much the World Cup as the gimcrack cups and trophies staged for the sake of television and sponsorship that left no trace save the money won and lost in the legal and illegal sports-betting industry. Cricket for the second time in its history faced a serious corruption crisis – one it has yet to overcome, for the reason that it is not an instance of a few bad apples in a barrel but a crummy bit of professionalised cooperage.

The shift in cricket's power balance has antagonised some. Cricket, it is complained, has been politicised; in truth, it was politicised the moment teams played under national banners. The disintegration of the Pax Britannica since the Second World War has merely brought to the surface a political dimension to the game previously obscured: no bad thing in itself, if the game's governors act with wisdom and in good faith. The shift, nonetheless, is replete with ironies and contradictions. The colonised have become the chief colonisers: it is the former outposts of empire that are now cricket's keenest missionaries. The more ostensibly democratic cricket's government, the more riven it has been by dissent: recently, cricket has lurched from controversy to controversy, whether the cause be bouncers, umpiring, sledging or ball-tampering. Spin has revolutionised cricket in the last decade, from the wrists of Shane Warne, Muttiah Muralitharan and Anil Kumble, and the public relations aides of administrators too numerous to mention. The problem is an abiding one. Because cricket has been for many countries the first important international sport, and thus a source of prestige and profile, it has tended to remain a captive of sovereign interests and self-seeking autocrats. Just because one dictator equals unilateralism does not mean that ten dictators equals multilateralism.

For a long time, a strong central government was seen as the answer to all cricket's woes. It was a false promise. The most effective government is still that inherent in the players and the fans: cricket retains an ingrained and unconscious sense of recognising what is not, of drawing its own bounds of the acceptable, of fair play. There has been, and is, much bad behaviour in cricket, much selfishness and truculence. But we remain capable of recognising the good, and attach a premium to it. It is not only for their individual brilliance that Sachin Tendulkar and Steve Waugh are admired: Tendulkar's serenity and Waugh's respect for tradition add to their stature, lengthen their shadows. That is why the drift towards police-and-punish discipline, absolving players of responsibility to the common weal and centralising power in an omnipotent ICC, is such a potentially destabilising departure from past practice.

Sport history is a paradoxical idea. Much of sport is about erasing history – of improving on, and thus eliminating, that which has gone before. Athletes often find the past a bane and a distraction. They'll stand on their records, yet dismiss those parts of them they regret or resent with a contemptuous: 'That's ancient history.' Cricket has a typically ambivalent relationship with its heritage: it has been a source of weakness and strength. Tradition and continuity have sometimes blinded it to social changes; they have also enabled it to withstand those changes, to hold modernity at bay long enough for adaptation to occur. In some ways, cricket has a history ideal for its needs, even in its mysteries. The bits we know are full of colour and character; the bits we do not imbue its emergence with a sense of the magical.

Wisden Asia December 2003

Victor Trumper
His Own Sweet Way

'To one we have read about, but never saw in action. He remains an immortal to all lovers of the great game.' With a bunch of red roses, a card thus inscribed was placed beneath a life-size bronze cast of Victor Trumper rendered by Louis Laumen as it awaited auction out front of the Melbourne office of Christie's a little more than three years ago.

The card was unsigned; the sentiment was universal. Though 2 November marks a century and a quarter since Trumper's birth, he maintains a subtle hold on Australian imagination: the cast eventually fetched a record price of $76,000. Sir Donald Bradman tops all polls as a symbol of national pride; Trumper has more romantic resonances. Sir Donald Bradman has been the subject of numberless books all of which manage to say the same thing; Trumper has featured in a single representation that is nonetheless polysemous.

The image in question is, of course, that captured by George Beldam, the Middlesex batsman and Edwardian photographer. Bernard Shaw once likened the photographer to a cod, needing to spawn in millions so that one offspring might make maturity; Beldam seems to have done rather better, producing from merely thousands of pictures offspring still going strong after a century.

The Trumper photograph was recognised at once as a classic. Beldam himself turned out 500 signed copies from a photogravure. It hung on the wall at Captain C.B. Fry's home at *TS Mercury* on the Hamble River, and on the hessian partition of a cottage occupied by Sydney urchin Arthur Mailey. For many years it featured on the cover of the New South Wales Cricket Association's annual report, and it remains the basis of the design adorning the cover of *Wisden Australia*.

It has been a motif for artistic enterprises as diverse as the Royal Shakespeare Company production of Graham Greene's *The Return of A.J. Raffles* and *Victor Trumper's Everlovin' Pop and Soul Revue* by Australian rock band You Am I five years ago. It has been turned into a print by Archibald Prize finalist Dave Thomas, a letterhead by Australian cricket bookseller Roger Page, the symbol of a chat room called the Victor Trumper Cricket Board, and will shortly feature on the jacket of *Endless Summer*, an anthology of writings about Australian cricket from *Wisden Cricketers' Almanack*.

At the simplest level, this is because Beldam's Trumper, in composition and content, is a very great photograph. It is both the first and last word in batting, insofar as batting consists of making instinctive what begin as a set of quite unnatural motions. In this image, Trumper seems to achieve the complete reconciliation of the orthodox and the spontaneous, the rehearsed and the original, the conscious and the unconscious. His shirt seems both loose and taut; his arms appear both gloriously free and tensed for action; the stroke is both instantly recognisable and uniquely his.

In *Great Cricketers: The Age of Grace and Trumper*, Beldam's son George Jnr suggests that the subject is intrinsic to the photograph's power: 'Would the result of this picture have been quite the same if the subject had not been the very man declared by all who saw him to be the most beautiful to watch, and whose charisma bewitched and enchanted even his own family of fellow-players? The answer must be: no.' Perhaps; perhaps not. This view may underestimate the enrichment of Trumper's legend by the picture itself. Beldam Jnr was right, though, to suggest that the photograph is of an era as well as of an individual. If it is possible for a cricketer to be their period, rather than merely be part of it, Trumper *is* the Golden Age of Cricket. In the gaiety and gallantry of his stroke play, the charm of his personality, even in his frailty, transience and suddenness of death, Trumper personifies what we understand as the values and nature of his time, both in absolute and comparative terms. Cardus awoke to what he regarded as the decadence of his own age while studying Beldam's Trumper in the upstairs tea-room at the Oval: 'A certain English batsman, vintage

1950, looked at this picture in my company and said, "Was he really any good?" "Why do you ask?" was my natural question. "Well," said this International, "just look where he is – stumped by yards if he misses." This sceptical England batsman had never in his life been so far out of his crease.'

There is, of course, more to Trumper than meets the eye in Beldam's image, and also to his period. The abandon of the stroke is not an altogether faithful representation of Edwardian cricket: it had its gross and greedy sides too. The Golden Age in Australia has also been particularly misread, being thought of as more or less akin to England's. As fine a writer as Mike Marqusee has been guilty of casually erroneous history, describing 'control of the game' in Australia as being a 'battle ... between Anglophile officials and the more "Australian" working-class players and spectators'. Which is the sort of analysis that gives Marxism a bad name. When it mattered, before the First World War, the contest was actually between two bourgeois castes: the players stood, more or less, for unfettered capitalism and entrepreneurship; the officials represented the forces of bureaucracy and centralisation.

Australian cricketers on tour in England were as interested in money as those today: they appointed their own manager, split the gate proceeds at tour's end, and pure *joie de vivre* was not the only explanation when they played entertaining cricket – it was also good box office. The difference to today was that Trumper's generation did not consider themselves professionals – professional cricket was an English phenomenon. In the midst of a wrangle between the players and the new Australian Board of Control in May 1907, an interviewer from the *Sydney Sun* specifically challenged Trumper with the description: 'The charge is that you are professional cricketers.' Trumper retorted: 'If we were, the inducements in England would soon take us there.' And Trumper spoke with authority, having declined an offer from Lancashire five years earlier.

So it is necessary, then, even though he is known primarily through a photograph taken by an Englishman in England, to understand Trumper in an Australian context. In Australian thinking, he became proverbial. When Australia's great general Brudenell White joined the British Army, official war historian Charles Bean lamented: 'We have lost our Trumper.' His name, as well as his reputation, was designed to live on. The first sailor to reach admiral's rank in the Royal Australian Navy was Victor Trumper Smith, born in 1913. It's not clear whether Trumper influenced the Bradman family's naming Don's brother Victor, but it would be a piquant link between the two paramount heroes of Australian batsmanship.

What he denoted to Australians, though, was not merely sporting hero, but self-made Australian sporting hero. As the historian Bede Nairn put it, Trumper 'reworked the charter of cricket from a Victorian artefact into an Edwardian palimpsest, with spacious Australian flourishes all but replacing the English script'. He connects with an Australian tradition of autodidacticism – one which subsequently also enfolded Bradman.

Trumper was tutored as a junior by Charles Bannerman, Test cricket's very first centurion, but proved uncoachable. Writing of Trumper's youth in *Town and Country Journal* in January 1913, S.H. Bowden recalled Bannerman's unheeded entreaties ('Leave it alone, Vic; that wasn't a ball to go at') and eventual decision to let the boy do as he pleased. The characteristic lasted. Monty Noble described it as Trumper's capacity for listening politely and attentively to all advice but going 'his own sweet way'. His immortality is not only manifested in the legend of his genius; it is reflected in the abiding Australian belief that individual flair and natural talent will out.

Wisden Cricket Monthly November 2002

Ranjitsinhji
The Indian Juggler

PERHAPS NO CRICKETER IN HISTORY has been as romanticised and sentimentalised as Ranjitsinhji. And it testifies to his uniqueness that this romance and sentimentality does not significantly enlarge on reality. W.G. Grace's prophecy that there would not be a batsman like Ranji for a hundred years is inherently untestable; but even assuming there has been, one could regard it as dependent on Ranji's pre-existence.

Ranji was, moreover, not only the first well-known Indian cricketer, but 'the first Indian of any kind to become universally known and popular' – as John Lord puts it in *The Maharajahs*. And this through a concatenation of circumstances that has inspired not only four biographies but two novels – Ian Buruma's *Playing the Game* and John Masters' *The Ravi Lancers*.

There was always a mythic element to 'Prince Ranji'. He was, most importantly, only fleetingly a real prince, adopted as heir to the Jam Saheb of Nawanagar purely as a precaution against that potentate's inability to father an heir of his own – a status revoked when this occurred four years later, and not restored for the next quarter of a century. In fact, this caprice was integral to Ranji's first exposure to cricket. Apprehension about court intrigues in Jamnagar resulted in his enrolment eighty kilometres away at Rajkot's Rajkumar College, the 'Indian Eton'; he developed there, under the influence of principal Chester Macnaghten, as a promising all-round athlete, before leaving in March 1888 for England to further his education.

Ranji's disinheritance, claims his most recent biographer Simon Wilde, also explains his determination at Cambridge to 'make himself the prince of cricket'. Indeed, the unique batting style that seemed to flow directly from Ranji's exotic origins belied the self-mastering dedication behind it. The tutelage of Surrey professionals who came to Fenner's to coach undergraduates was most valuable. Ranji reputedly tackled them in relays. 'I must practise endurance,' he told F.S. Jackson. 'I find it difficult to go on after thirty minutes.'

The legend is that one of these professionals, to cure the young Indian's tendency to retreat from the ball, nailed Ranji's back boot to the crease. And true or not, the facility that Ranji subsequently developed for deflecting balls to leg inflected the course of batsmanship forever, opening quadrants of the field previously unexplored and untenanted. Cricket, hitherto, had been an off-side game. Ranji's great friend C.B. Fry recalled of his instruction at Repton: 'If one hit the ball in an unexpected direction on the on side, intentionally or otherwise, one apologised to the bowler ... The opposing captain never by any chance put a fieldsman there; he expected you to drive on the off-side like a gentleman, even if his bowlers presented stuff which ... one could turn into long-hops to leg.' Ranji's 1897 *Jubilee Book of Cricket* illustrated this occidentalism in diagrams of fifteen recommended fields for bowlers of various types: only one, for 'lob bowlers', contains more than three leg-side fielders, and most feature but one or two. Ironically, more than any other batsman, Ranji rendered such formations obsolete.

Ranji's harnessing of the pace of the ball in the power of the stroke was also revelatory; as Gerald Brodribb contended in *Next Man In*, it 'suggested a completely new way of getting runs'. With wrist, eye and timing, batting became a discipline of touch and subtlety as much as strength and force. And not even a modern batsman would make so bold as Ranji, if one judges by Fry's description in *Batsmanship* of his contemporary's 'forward-glance-to-leg': 'The peculiarity ... was that he did more than advance his left foot at the ball ... he advanced it well clear of the line of the ball to the right of it, and he somehow got his bat to the left of his left leg in line with the ball, and finished the stroke with a supple turn at the waist and an extraordinarily supple follow-through of wrist-work.'

How Ranji's colour hampered his advance is difficult to ascertain at this remove. There was prejudice at Trinity College against 'a nigger showing us how to play cricket', and Sir Home Gordon revealed in *Background of Cricket* that Ranji's signature habit of buttoning his sleeves to the wrist was 'acquired at Cambridge to mitigate his dusky appearance'. Gordon also recalls that Ranji's selection for the 1896 Ashes series was debated at Lord's even after his stupendous Old Trafford debut (62 and 154 not out): 'Old gentlemen waxing plethoric declared that if England could not win without resorting to the assistance of coloured individuals of Asiatic extraction, it had better devote its skill to marbles. Feelings grew so acrimonious as to sever lifelong friendships ... one veteran told me that if it were possible he would have me expelled from MCC for having "the disgusting degeneracy to praise a dirty black".'

Yet it is remarkable, as Alan Ross commented in *Ranji*, how quickly he 'became a "character"'. By the end of Ranji's debut series, the ramparts of Victorian England had yielded to him utterly. 'At the present time,' judged *The Strand*, 'it would be difficult to discover a more popular player throughout the length and breadth of the Empire.' And his fame would only grow: he published a string of books, was the subject of a biography by Percy Cross Standing and a painting by Henry Tuke, edited an edition of *The Sun* and appeared in *Finnegans Wake* as a pun.

The time was ripe for a crowd-pleaser of Ranji's kind. With county cricket's blossoming, he made the most of the expanding first-class program: between 1898 and 1901, he and Fry, a combination unmatched in glamour, amassed 16,500 runs for Sussex. Standing noted that visitors sometimes left disappointed at Ranji's indifference to cricket past; but 'what's he to Fuller Pilch, or Fuller Pilch to him? Nothing. On the other hand, today's weather forecast is everything.' Perhaps he intuited that history was of little account, save for that he might make.

Ranji's, though, is an era usually identified with imperialism and white supremacism. Why was his passage relatively frictionless? In his *Ornamentalism: How the British Saw Their Empire*, David Cannadine contends that rank, as much if not more than race, was the key to imperial order; the empire's proprietors identified with the aristocracies of colonised countries rather than merely subjugating them. This sprang, he contends, from social and political conservatism, the perception that British institutions at home were imperilled by vulgarising capitalism and democracy: 'India's was a hierarchy that became the more alluring because it seemed to represent an ordering of society … that perpetuated overseas something important that was increasingly under threat in Britain.'

The crown's assumption of direct responsibility for the administration of India after the Mutiny not only recognised royal courts like Ranji's Nawanagar but celebrated them. Debating the Royal Titles Bill, Benjamin Disraeli proclaimed that India's princes 'occupy thrones which were filled by their ancestors when England was a Roman province'. 'From one perspective, the British may indeed have seen the people of their empire as alien, as other, as beneath them,' says Cannadine. 'But from another, they also saw them as similar, as analogous, as equal, and sometimes even as better than they were themselves.'

Ranji, *soi-disant* 'prince', famous for his ostentatious regalia and improvident flourishes of wealth, colluded in those fantasies of an imagined east. On one occasion when they dined at Hyde Park Hotel, Gordon was struck by the shabbiness of Ranji's suit, but noted at a military tournament later in the evening that Ranji took the salute in the royal box 'resplendent in Oriental attire and ablaze with jewellery'. Appreciations of Ranji's batting, too, are often from a perspective of inferiority, even of gratitude. 'We are a conservative nation, our chief aim being to imitate our grandfathers,' wrote Tom Hayward in his 1907 primer *Cricket*. 'What was good enough for them we have considered good enough for ourselves, and, truth to tell, we had got into a groove out of which the daring of a revolutionary alone could move us. The Indian Prince has proven himself an innovator. He recognises no teaching which is not progressive, and frankly he has tilted, by his play, at our stereotyped creations.'

By this time, Ranji had essentially renounced cricket, his political ambitions having fructified with the restoration of his inheritance after the death of the Jam Saheb in August 1906. But Neville Cardus' remark that 'when Ranji passed out of cricket a wonder and a glory departed from the game forever' was hyperbolic: cricket's inheritance from him is enduring.

Wisden Cricket Monthly March 2002

Sir Jack Hobbs
The Master

AFTER HIS RECENT MONOPEDAL HUNDRED in the Oval Test, Steve Waugh was asked by Channel Four to explain his overpowering desire to participate in the game. He offered a succinct and direct response: 'I'm a professional cricketer and I love playing for Australia.'

It was a typical Waughism. Yet at one time, the remark would have sounded paradoxical, even dangerous. There was cricket for pleasure and cricket for profit, amateur and professional, oil and water. Then it was commingled, most successfully, sympathetically and far-reachingly, by Sir Jack Hobbs: in the words of Sir Pelham Warner, as though describing one of the marvels of the age, 'a professional who bats exactly like an amateur'.

Last year, Hobbs was garlanded as one of *Wisden*'s Five Cricketers of the Century: the only Englishman. His status as the greatest first-class run-scorer (61,237) and century-maker (197) clearly counted for something; his serene and sportsmanlike demeanour for something more. Yet seventy respondents to the survey did not figure him in their calculations, while Matthew Engel's *Almanack* tribute seemed somewhat tepid: Hobbs was called 'pragmatic', 'businesslike' and 'the supreme craftsman', though 'not an artist'. And some critics booed the choice; one even chided me in print about using 'the c-word' – 'class' – in explaining his eminence.

This dubiousness, I suspect, has more to do with us than with Hobbs. Now that 'fings ain't what they used to be' has been swept away by 'fings've never bin better', anything old now seems frightfully staid and inhibited. But was Hobbs really like that? Study the images composing that much-neglected 1926 primer *The Perfect Batsman*, and one obtains a different impression. This book's '98 Cinema-Photographs of J.B. Hobbs at the wicket' were taken for author Archie MacLaren by Messrs Cherry Kearton Ltd in 1914, at which time Hobbs felt himself at his peak. And they are anything but staid, or even conventional, and not a bit 'pragmatic': the bat speed and *brio* are breathtaking. In *Jack Hobbs*, John Arlott remarks on his subject's tight bottom-hand grip ('contrary to the advice of most coaches'), and it is evident in the sequences that illustrate 'Driving to the Right of Cover Point' and 'A True Cover Drive Along the Ground'. It is batting at its most spontaneous and original; 'The Master' and 'The Master Blaster' were not quite so distant as might be imagined. Indeed, Hampshire's Alec Kennedy once recalled bowling the first ball of a match to Hobbs at the Oval: Hobbs despatched a late outswinger on off stump through square leg for four. The anecdote's only un-Vivish aspect is that Hobbs smilingly apologised: 'I shouldn't have done that, should I? I was a bit lucky.'

To appreciate Hobbs' full significance, however, we must grapple with that troublesome 'c-word'. In the cricket of Hobbs' time, class is what it was all about. The discrimination that placed amateur and professional in separate dressing rooms and separate hotels, and that saw them enter the field by different gates and travel in different railway carriages, now seems as remote from our quotidian experience of cricket as sectarianism or McCarthyism. Yet it was very real, and even seemed the natural order; not least to Hobbs, who enumerated his Surrey colleagues thus: 'There was Tom Shepherd, Andy Ducat, Mr Jardine, Mr Jeacocke, Mr Fender, Bob Gregory, as well as Andy [Sandham].'

Quietly but categorically, Hobbs upset that precedence. Before him, professionals had been largely chained to the bowling crease. Only a handful of 'players' had represented England as batsmen: Arthur Shrewsbury, Bobby Abel, Johnny Tyldesley, Tom Hayward. And these were business cricketers; when Hayward received merely a fiver in talent money for scoring 315 not out at the Oval, he remarked: 'Well, it's no use me getting 300 again.' Heavily influenced by Hayward, Hobbs was clear-eyed about his vocation: 'Unless you get to the top where the plums are, it is a bare living, and when your cricket days are over, you have to find a new career.' But he bucked the stereo-type of the workman pro, for commercial realities never restrained his adventure or compromised the pleasure he took in his craft. 'Of course I was earning my living,' he said. 'But it was batting I enjoyed.'

Hobbs was not an agitator for the rights of the professional; rather the excellence of his example provided a personification of the cause. His Surrey skipper Percy Fender was one of the first to lead his team on to the field through a single gate – earning the reproach from Lord Harris: 'We do not want that sort of thing at Lord's, Fender.' It was because of Hobbs that Fender felt free to advocate the unthinkable notion of professional captaincy: 'My experience of professional crick-eters does not teach me that cricket would necessarily go to the dogs if a professional happened to be in charge.' And it was Hobbs that Cecil Parkin invoked as the best leader he'd played under in his famous January 1925 article in the *Weekly Dispatch* disputing Arthur Gilligan's Test captaincy credentials, provoking Lord Hawke's even more famous retort: 'Pray God that no professional will ever captain England.'

The response to that plaint was the greatest testimony to Hobbs' renown; even the former Australian governor-general Lord Forster weighed in, stating that he would 'never hesitate to play under the captaincy of a man like Jack Hobbs'. And so it came to pass that on 26 July 1926, Hobbs was asked to lead England when Arthur Carr fell ill during an Ashes Test at Old Trafford: the first professional to do so since Shrewsbury (who had, it should be pointed out, organised his own tour). Hobbs was reticent. He disliked the responsibility of lead-ership, and his public position was that 'most professionals' preferred an amateur skipper; but he took the job, drew the Test, and foresaw others following in his footsteps. 'We are such sticklers for tradition in insisting on an amateur captain, regardless of the question whether he can pull his weight as a player or not,' he wrote in *My Life Story*. 'The time is coming when we will have to change our views ... when there will be no amateurs of sufficient ability to put into an England side.'

Cricket, as so often, found ways to defer change. And as a stan-dard bearer for the exploited, Hobbs would not appeal to a Marxist:

he overthrew nothing, stormed no citadel. Yet, in a way, he so digni-
fied his profession that the archaisms of class distinction in cricket
had ceased to matter long before the 1963 annulment of the 'gentle-
man/player' divide; Alec Bedser commented that using the players'
gate at the Oval was no hardship because, if it had been good enough
for Hobbs, it was good enough for him. That 'Sir Jack' became the
first professional games player to be knighted in 1953 is one of cricket
history's better coincidences: England was in the course of winning
the Ashes under its first full-time, externally appointed professional
captain, Len Hutton.

If all this seems like history – and there's no remark in the contemporary sporting vernacular more dismissive than 'that's history' –
Hobbs' career had two other trappings of modernity that loosely link
him with Steve Waugh. The uxorious Hobbs was the first professional
to take his wife on an Ashes tour. He did it characteristically, initially
declining to make the 1924–25 trip, then agreeing to join the team as
an 'extra member' when Lord Harris granted Ada Hobbs permission
to accompany her husband. Waugh has taken up a similar cudgel,
championing a new Australian regime that welcomes players' partners on tour.

Likewise does Waugh follow in Hobbs' literary path, capitalising
his fame through the publication of books. Nine books bear Hobbs'
name. Eight bear Waugh's. This is doubtless another record that
Australia's captain will set himself to breaking.

Wisden Asia December 2001

Jack Gregory
The Mighty Atom

THE CAMERA LOVED JACK GREGORY. Photographers of his day had still to conquer the distance between boundary and pitch, but Gregory's expansive, vibrant cricket reached out to meet them, advancing, radiating from the middle.

There's a famous image of him in his delivery while practising at Lord's: the front leg hanging in mid-air, the right arm at the bottom of its swing, the left arm aloft. The body's slight backward tilt lends him a palpable energy. Another picture features him batting at Sydney a few years later. With the completion of a drive, the hands have ended up over the right shoulder. The force of the stroke has turned Gregory's trunk to face down the wicket, although the ball is disappearing through mid-off, and the front foot hovers just off the ground. Here again is energy, a glad, spontaneous vitality. To enhance the breezy naturalness, Gregory is bare-armed, bareheaded and gloveless; indeed, we have it on Ray Robinson's authority that he even scorned a protector.

Not so long ago, I chanced on another photograph I'd never seen before, on the front page of a copy of Sydney's *Referee*. It was Gregory taking a slips catch, and a blinder, too: parallel to the ground, the ball snug in an outstretched right hand, while wicket-keeper Hanson Carter cast a startled glance over his right shoulder. More than seventy years have elapsed since Jack Gregory played cricket – yet it was hard to avert my eyes. Eighty years since his first breathtaking feats, age hasn't wearied him; he strides out of the past as the most magnetic and expressive player of his day.

Donald Bradman, incidentally, thought the world of 'vital and vehement' Gregory. 'His bowling in the early days was positively violent in its intensity,' opines the Don in *Farewell to Cricket*. 'His whole attitude towards the game was so dynamic, his slip fielding so sensational, his brilliant batting so pleasing … that thousands would flock to see Gregory play anywhere at any time.'

Another who 'venerated' Gregory was, oddly enough, Harold Larwood. 'Gregory's bowling was the essence of savagery,' wrote Bradman's chief nemesis, 'his great kangaroo leap at the last instant presenting an awesome sight to any batsman.' It is surely something to have both Bradman and Larwood in a corner on any subject, yet admiration of Gregory united them.

*

An arresting feature of Gregory's career is that it might easily never have happened. Father Charles had played for New South Wales, uncles Dave and Ned for Australia, the former as captain in the inaugural Test match at Melbourne in March 1877, the latter becoming the architect for the Sydney Cricket Ground's famous scoreboard (disused today, it still peers furtively over the brow of the old Hill). Another uncle, Albert, was renowned for owning Australia's most comprehensive cricket library. And cousin Syd had, ten months before Jack's birth, relieved England of the first Test double century on Australian soil.

Yet Jack the Lad evinced no particular aptitude for cricket. He captained Shore's first XI in 1911 and 1912, but as a batsman, and impressed at least as much as a hurdler and first XV rugby player. On leaving school, he went jackerooing in Queensland; a source, he wrote later, of deep disappointment to his family. By twenty, he cut a striking figure: six feet three inches tall, bronzed, blue-eyed and rugged. In portrait photographs, he has an air of alertness bordering on impatience, as though in a hurry to be elsewhere. But it was not to be playing cricket.

Cricket's prospects of claiming Gregory dwindled further when he enlisted as an artillery gunner in January 1916. He served in the 7th, 10th, 11th and 23rd Field Artillery Brigades and 3rd and 4th Division Artillery Brigades, undertaking two tours of France. He played regimental cricket, and there first toyed with fast bowling: the legendary Charlie Macartney first met Gregory in an inter-unit game on a matting pitch at Larkhill, Salisbury Plain, and remembered him for ending the match by hitting a batsman in the eye with a lifter. But otherwise Gregory's off-duty recreational energies were devoted to winning the hurdles over 120 yards, the sprint over 100, and his unit's tennis championship.

With peace, however, came the notion, mooted by the MCC grandee Pelham Warner, of a touring cricket team selected from AIF ranks; so much of the services' recruiting effort had been aimed at sportsmen that it was heavily endowed with cricketers, and a number of unofficial internationals had already been staged between England and an ersatz Dominions XI. Players were invited to join training sessions preparatory to selection at the Oval; Lieutenant J.M. Gregory was one.

Quite why Gregory should have featured among invitees is unclear. Before enlisting, he'd played exactly one game in Sydney first-grade. According to historian Ron Cardwell, the summons was on the recommendation of another officer in Gregory's unit, former New South Wales Sheffield Shield batsman Frank Buckle. Gregory heard that his invitation came direct from Warner, who'd watched him play in a social match in which Gregory had represented the Artillery Officers' School against Red Cross, and surmised latent ability from the family name. In the event, the twenty-four-year-old was the last of the touring sixteen picked, and then largely on the basis of physique; if he couldn't play much, it looked like Gregory could learn.

In hindsight, the AIF team was a powerful one. Lance-Corporal Herbie Collins, who became its captain, went on to lead Australia. Corporal Bert Oldfield, Captain Clarence Pellew and Gunner Johnny Taylor would serve under him as Test men. Then there was Gregory himself, all his cricket ahead. Yet at the time, their quality was unknown, their resilience unascertainable. Some had endured very hard wars indeed. Oldfield had spent five months in a military hospital with shell shock. Captain Bill Trenerry had been wounded twice and Lieutenant Charles Kellaway no fewer than four times. Gregory himself had lost a cousin at Gallipoli. Their financial incentive, more-over, was modest indeed: 8s a day allowance on top of their military pay.

After scraping into the team, Gregory advanced by accident. Early on, he reported having cut his hand by stepping on it at fine leg. He was ribbed mercilessly; he was nicknamed 'Pavlova', an ironic refer-ence to the dancer, while Collins commented that he had all the elegance of a tank rolling into action and ruefully directed him to con-valesce at slip. To universal surprise, however, 'the long 'un' proved a close catcher of uncanny reflex and reach; scorer Bill Ferguson likened his arms to an octopus' tentacles, 'only they seemed twice as many and twice as long'. Impressed, Collins tossed him the ball when the leader of the team's attack, Captain Cyril Docker, broke down. It was like removing a genie from a bottle. In the tour's third match, against Cambridge University at Fenner's, Gregory claimed 6/68, and in bowl-ing two batsmen didn't simply disarrange the stumps but knocked them flying like ninepins. On a benign surface at Lord's a few days later, he struck the chest of Middlesex's Mordaunt Doll so hard that the ball lifted the batsman off his feet, before trickling cruelly onto the stumps. Doll departed. Gregory had arrived.

*

Gregory would bag 178 wickets at less than 17 on that trip, and collect 1352 runs at 31. But a recitation of his successes wouldn't convey the force of his initial impact. Here was a mediocre grade cricketer, three years at war, suddenly transformed into a self-taught fast bowler, hitter and catcher. And he was dramatic. Oldfield described his run as 'ungainly', for his stride was immense and his boots so voluminous that team-mates could fit their own shod feet inside them, but it was also unsettling, for what would result was always uncertain, perhaps even for the perpetrator. The action climaxed in what soon became known as a 'kangaroo hop' in the delivery stride, and was completed with a follow-through that rolled like a tsunami halfway down the pitch. And the effect of Gregory's hyperkinetic exertion was fearful. Writing fifty-five years later in *Wisden*, Sir Neville Cardus insisted that

Gregory 'ran some twenty yards to release the ball'. In fact, Gregory's approach was only fifteen yards, twelve paces; it simply 'seemed' longer.

As the ball descended from almost eight feet, it could do almost anything. When he pitched up, Oldfield thought, Gregory swung the new ball as much as any bowler he ever took. When he pitched short, Collins remembered, the ball seemed to whoosh like a mortar. For he was unrelievedly, inexhaustibly quick. Wrecking the stumps of a South African batsman called Phillip Hand in Johannesburg on the AIF tour's homeward leg, Gregory sent a bail flying forty-six yards.

Gregory's batting, initially primitive, always uninhibited, also developed on that AIF tour. He scored his maiden first-class hundred against Northants, and was eventually trusted to open the innings. Then there was the fielding, infallible at slip, electric anywhere. Against Natal at Durban on the first three days of November 1919, Gregory harvested 9/32 and threw the last man out for good measure.

When the AIF team landed in Melbourne two months later, Gregory was the object of insatiable curiosity. Games were arranged to test the mettle of Collins' now-celebrated cohorts. At his first Australian gallop, Gregory claimed 7/22 against Victoria, bowling flat out on a rain-affected Melbourne surface. A fortnight later against New South Wales, opening both batting and bowling on the arena adorned by his uncle's scoreboard, he scored 122 and 102, took eight wickets and three catches. In eighty years since, no one has even approximated such a feat, let alone paralleled it.

Nowadays, Gregory would be fending off media and public alike. In fact, discharged in March 1920, he returned to the land he had forsaken, working again as a jackeroo in Dubbo for a pastoralist, Fred Body. The only time he exhibited his prowess over the next five months was when the property suffered a mouse plague, so severe that the homestead's wire doors were literally choked with vermin. Gregory would entertain his fellows while they waited for this flood tide to abate. The jackeroos would quickly open the doors, and Gregory would catch the jumping mice with either hand – he never missed.

There was no doubt, however, that he would be needed again. Johnny Douglas's English team, which toured the antipodes the following season, must have sensed what it was in for. And, in 1920–21, Australian cricket had a pitiless edge. New captain Warwick Armstrong was a leader who took the tact out of tactics and, with home Tests of the time played to a result, beaten teams had nowhere to hide. The hosts won all five contests and, if Gregory did not completely dominate the rubber, that was only because Armstrong had such a constellation of talent at his disposal: the left/right opening alliance of Collins and Bardsley, the hawk-eyed Macartney and the googly guru Arthur Mailey among them.

As it was, Gregory would almost certainly have been man-of-the-series, had such awards been in vogue. No Australian has approached his 7/69 and chanceless 100 in 137 minutes in the Second Test at Melbourne over New Year. For good measure, he also intercepted no fewer than fifteen chances. Sir Jack Hobbs said Gregory stood closer than any slips catcher of the time, and was unmistakably ravenous for the ball; sometimes, when Mailey loosed his wrong 'un, Gregory would take a couple of giant strides from his normal post and materialise, like an apparition, at leg slip.

Gregory formed two vital strategic alliances during this series. The first was with his captain. The prickly Armstrong was a figure more respected than admired by contemporaries; sooner or later, he seems to have antagonised most opponents and more than a few comrades. Yet Gregory revered the vast Victorian, referring to him in two very rare articles some years later for *Sporting Globe* as 'my ideal cricketer' and 'the greatest of my time'.

Gregory also aligned with another cricketer from Victoria, although originally a tinsmith's son from Launceston, Ted McDonald. Like Gregory, McDonald had begun cricket as a batsman, before discovering a natural gift for bowling fast; the great Englishman Frank Woolley thought him speedier even than Larwood. As a combination, they were fire and ice. Where Gregory's approach shook the earth, McDonald's flowed like a river. While the former enjoyed the wind at his back and swung the ball mostly away, the latter didn't mind breasting a breeze and possessed a sinister breakback. To Cardus, Gregory was 'atomic-powered', McDonald 'Satanic'. On one key question, however, Gregory and McDonald were as one. As McDonald's future Lancashire team-mate Cec Parkin put it: 'He always maintains … that there never was a captain to equal Warwick Armstrong.'

McDonald actually achieved little in the three Tests of that 1920–21 Ashes series, and had reason for gratitude to Armstrong at its conclusion; the captain insisted on his selection to partner Gregory for the forthcoming tour of England. It was a groundbreaking recommendation. Hitherto, the concept of entrusting the new ball to pace at both ends had been almost unknown; the great Australian pacemen of yore had been solo venturers (Fred Spofforth, Ernie Jones, Tibby Cotter), usually sharing the new ball with a medium-pace bowler or even a spinner. With Gregory and McDonald, speed was now in the ascendant; now, in fact, and evermore.

*

Neither Gregory nor McDonald fit the contemporary fast bowler's mould of histrionic aggression, and of a general love of attention. Gregory was anything but a limelighter. As Armstrong's Australians crossed the continent before boarding the England-bound *Osterly* in Fremantle in March 1921, the train had to stop in country South Australia for a civic reception at Quorn. As manager Syd Smith addressed the crowd, a cry went up for a word from Gregory and a posse of locals boarded the train in search of him. The first player they encountered was Gregory himself but, in those days when news was vested in word rather than image, he went unrecognised. Thinking quickly, Gregory confided that the man they sought had slipped off the other side of the train, sending his admirers off in comic pursuit.

Bradman noticed that Gregory could never be drawn to discussing his on-field accomplishments, though unfailingly 'generous-hearted' towards younger team-mates. Larwood, so exhilarated by Gregory's bowling, also found him the gentlest of men away from the fray: 'Off the field you could not meet a more friendly and amiable chap.' Lecry of the press after once being misquoted, the only interview he granted was a most reluctant one in the year before his death, an aspiring cricket writer called David Frith driving 320 kilometres on the off-chance of catching him in his shack at Narooma on the New South Wales south coast, and coaxing him into a few quiet reminiscences. There were no visible trophies or mementoes. 'Here,' recalled Frith, 'was a cricketing Garbo.'

McDonald, likewise, was a self-contained character with little small talk, among strangers as inscrutable as a cigar-store Indian; Ronald Mason thought that he 'harboured in a not very approachable personality a genuine vein of genius'. His energy seemed almost supernatural. Even bowling at his fiercest, Oldfield recalled, he did not seem to perspire. On one occasion, he arrived late for a day's play, drained a glass of water, took a couple of drags on a cigarette, and bowled unchanged till lunch. Yet that summer of 1921, these quiet assassins sapped English spirit where bombs dropped on London from the Kaiser's Zeppelins had failed. In the first innings of the First Test at Trent Bridge, Gregory had Donald Knight caught behind then bowled Ernest Tyldesley and Patsy Hendren in one hectic, breathless over. In the second innings, he bowled Tyldesley again, this time off his head. No Man's Land had been recreated over twenty-two yards of turf.

For the Second Test at Lord's, England picked a batsman called A.J. Evans: a good player, and a courageous man. As a pilot during the Great War, he'd won the Military Cross and bar. As a POW, he'd organised escape attempts of such reckless derring-do that he'd been persuaded to pen a book, *The Escaping Club*, describing how his mother had sent him maps and compasses concealed inside cakes and jars of anchovy paste. But on the day of his Test debut, a team-mate remembered: 'He was so nervous that he could hardly hold his bat, and his knees were literally knocking together … His nerve had gone and the first straight ball was enough for him.' Australia won those Tests, and the next at Leeds, to retain the Ashes, and against the counties waged a reign of terror. London's *Sun* published a page of diagrams of Gregory's fields headed 'Waiting for the Express'. The wait itself was fearful. Even Wally Hammond, then a rising star with Gloucestershire on the brink of a brilliant thirty-year career, had not the stomach for it: 'Jack Gregory had cultivated a fearsome stare and gave me the treatment. With knees trembling and hands shaking, I was relieved when he bowled me first ball.' Major Philip Trevor, the *Daily Telegraph*'s long-serving cricket correspondent, found the implications of Gregory's dominance rather uncomfortable: 'He has taken advantage of an era in English batting – a rather deplorable era I think – in which a blend of moral courage and physical courage does not count as high as it did in the days when the Grand Old Man "picked 'em out of his beard" and Maurice Read and others had no more tender regard for their necks and shoulders than a good rugby forward.'

It was Armstrong's turn for gratitude. No Australian touring party has won more first-class games. Gregory and McDonald claimed 116 and 138 wickets in England respectively, each at the pin's fee of 16.6 each. In twenty-seven matches, Gregory also accumulated 1135 runs at 36.6 and hauled in 37 catches; no Australian all-rounder has performed such a 'double' since. En route home via South Africa, he then annexed the honour of the fastest Test century, belting 34 from his first twelve deliveries, and cruising to three figures in ten minutes over the hour. Those who regard packed international schedules as a feature only of the jet age should inspect Jack Gregory's figures in the first three years of his cricket career: eighty first-class matches on three continents for more than 4000 runs at almost 40, and 400 wickets at 17.

*

Perhaps it was too much. On his return home, Gregory foreshadowed retirement, and his intention of returning to the rural life; his right knee was by now constantly painful and inflamed. In the event, he found a job with a Sydney sheet-metal manufacturer, Kavanagh & English, and submitted to a surgeon removing a cartilage from the afflicted joint: an operation at the time from which recovery was never total. Indeed, despite being an absentee from the next two domestic seasons, Gregory was seldom the same. On occasion, he offered glimpses of his former threat, once breaking the shoulder of Arthur Richardson's bat in a Shield match at Adelaide Oval as he had him taken at slip. More often, he was merely a good bowler rather than a great one.

By coincidence, his admirers Larwood and Bradman were both present when Gregory's knee finally buckled during the Brisbane Test of December 1928: Larwood batting in his first Australian international, Bradman making his Test debut. As Gregory dived headlong for a return catch, his knee bore the brunt of the fall. Larwood recalled the Australian's tears 'not from the pain of the injury, but from the realisation that he could no longer play the game he loved'. Bradman recorded Gregory's words when he was carried to the dressing room: 'Boys, I'm through, I have played my last game.'

A photograph exists of the moment Gregory was maimed; as I said, cameras were seldom idle in his presence. Two features compel attention. Larwood's bat has shattered at the shoulder; a final attestation of Gregory's force. And Gregory's prone figure lies more than halfway down the pitch; a last intimation of his athleticism. Jack Gregory was right; he never played again. But, at a glance, he continues to come alive.

Test Team of the Century (2000)

George Headley
Atlas

GREAT CRICKETERS ARE USUALLY best contemplated at the Test
arenas in which their great deeds were done. But to appreciate
George Headley, it's probably better to visit a club game, in a park, on
a village green, or a maidan, looking for that incongruous presence:
the really good cricketer in a field of modest triers.

Everyone has watched or played with someone of the kind. He
seems in synch with the game, to understand it, a natural compared
with whom team-mates look ungainly. He usually goes in number
three and bats all day, husbanding the strike. When he succeeds, his
pals are in good heart. When occasionally he fails, victim of an out-
landish catch or a bad call, they mount little further resistance.

For a decade of Test cricket, this mantle was George Headley's.
To evoke the batting burden he bore for the West Indies, C.B. Fry
famously dubbed him 'Atlas'. Actually, the classical allusion isn't
quite exact. Unlike the mythological Atlas, lumbered with a job he
detested and hoodwinked by Hercules into taking it back, Headley
never shirked his burden. But Fry might also have been recalling his
Homer, who made Atlas father of Calypso.

The complications of a role such as Headley's must be understood
as the most formidable in cricket. Great players usually play in good
teams. This is a reason they become great, because they have the
opportunity to bat with competent partners and in favourable circumstances, or to bowl with reliable backup and alert fieldsmen. The
lot of the outstanding player in a mediocre team is disproportionately
harder, not merely because of the absence of able support and the
likelihood of losing causes, but because the scenarios encountered
tend over time to distort one's natural game.

For a batsman, questions arise. Do you preserve your wicket at all costs in the hope of prolonging a game? Or do you risk all in the knowledge that time is short? And as explained of his own career by John Reid – whom John Arlott once described as 'an Atlas-type figure of New Zealand cricket' – you must in a sense ignore reality: 'I told a lot of lies. We'd gather as a team and naturally I'd try to be as positive as possible … I'd try to encourage our fellows, to explain that everyone is human, that they all got nervous, had failures. But in the back of your mind there was this knowledge that all things being equal, we were in for a rough time.'

Headley dawned, furthermore, ahead of cricket's institutional structures in the West Indies. His first series was in 1929–30, only two years after the foundation of the West Indies Cricket Board of Control. The region had no domestic competition, and administration was still radically decentralised: each venue had a different selection panel, chose a different captain, and in all twenty-eight players in the four-Test rubber against England. Twenty-year-old Headley, indeed, was the only Jamaican summoned for the Tests until Kingston hosted the last: not surprisingly, in view of his 703 runs at 87.9.

For the next decade, Headley was first West Indian picked everywhere, the other ten players almost an irrelevance. In nineteen prewar Tests, he made 26.9 per cent of West Indian runs – a greater ratio even than Bradman, who made 26.5 per cent of Australian runs in his thirty-seven pre-war Tests. Headley scored two-thirds (ten of fifteen) of the West Indian hundreds in his appearances; Bradman not quite half (twenty-one of forty-six) of the Australian hundreds in his.

C.L.R. James advanced Headley as 'my candidate for a clinical study of a great batsman as a unique type of human being both mentally and physically'. When team-mates fell to foolish shots, he would simply ask, as if puzzled: 'Why him don't like to bat?' For his own slight build, round shoulders and plod to the crease with bat dragged behind him belied a batting personality that was instinctively dominant. Jeff Stollmeyer recalled that Headley was fond of smacking the first ball from a spinner straight back, trying to hurt their hands: 'It was George's way of ensuring that spin bowlers did not

34

give him much trouble.' Headley's placement was so precise, Learie Constantine remembered, that he would pick out fielders who were also bowlers, endeavouring to tire them: 'Sometimes he places the ball with fiendish cunning, so close and tempting that the player strains a shade too much to make an impossible catch or stop a ball a foot beyond his outstretched finger tips; and then a muscle is pulled or an ankle dragged.'

At a time when West Indian cricketers were stereotyped as ebullient and undisciplined, Headley was also a pragmatist. He once recommended that the West Indies bowl outside leg stump in order to deny India a Test victory; it worked. His record bespeaks uncommon drive and endurance. Headley's average at the outbreak of war was 66.7; among team-mates who played as many as five Tests, the next highest average was 30.7. West Indian captains in that period, meanwhile, had contributed only seven half-centuries – but they, of course, were white.

Headley became, as former Jamaican prime minister Michael Manley put it, 'black excellence personified in a white world and a white sport'. He was proud of his Afro-Caribbean heritage, describing himself as 'African' on the immigration form when he entered Australia for the 1930–31 tour, and occasioned pride in others. On seeing a photograph of a line of cricketers waiting to meet George VI, historian Frank Birbalsingh felt uplifted: 'That one of us – a black man – could shake the hand of a king introduced possibilities formerly undreamt of in our colonial backwater of racial inferiority, psychological subordination and political powerlessness.' But being fit for kings did not, in the eyes of the WICBC plantocracy, fit Headley himself for leadership. As Constantine wrote: 'Cricket in the West Indies is the most glaring example of the black man being kept in his place ... The heart of our cricket is rotted by racist politics. I only hope that before I die, I see a West Indian cricket team chosen on merit alone, and captained by a black man, win a rubber against England.'

Constantine almost had his wish in January 1948, when Noel Nethersole, deputy of Jamaica's People's National Party to Michael

Manley's father Norman, agitated from within the WICBC for Headley's recognition. But Headley's tenure was initially confined to the First and Fourth Tests against Gubby Allen's visiting Englishmen, a gesture reeking of political expedient. From the outside, Wally Hammond expressed bafflement: 'Headley was by far the outstanding player as well as the most experienced cricketer ... and I do not see why he was not given unqualified control of the team for the whole series.' From the inside, Clyde Walcott described widespread belief that the WICBC, 'while realising that he [Headley] was the right man for the job and not daring to usurp his rightful place on his home ground of Jamaica, felt much more confident of their powers in Trinidad and British Guiana'.

In the end, a back injury that Headley sustained in the First Test kept him from the last three Tests, and the captaincy settled on white John Goddard. Another twelve years elapsed before Frank Worrell's appointment fully enfranchised black West Indian cricketers – and by then, Worrell could call on immeasurably stronger XIs. Historian Hilary Beckles calls Headley 'a saturnalia – a sign of things to come'. Like the epithet Atlas, this is both apt and not quite apt: great cricketers would indeed follow Headley's footsteps, but none would again have to bear the responsibility of being their team's first, best and only hope.

Wisden Cricket Monthly June 2002

Bradman's 254
Cracking Noises

SIR DONALD BRADMAN'S CAREER INVOLVES many totemic numbers. In a cricket publication it is almost superfluous to mention the contexts in which 99.94, 6996, 334 or 974 arise. But no number resonates quite like 254, because no innings ended up holding for its maker quite the same significance.

The first of Bradman's great Ashes innings, at Lord's in June 1930, was like his very own Operation Shock and Awe, with English cricket's dismay as its objective. It commenced at 3.30 p.m. on the second day of the Second Test with what remained Bradman's fastest Test fifty, in forty-five minutes. What Neville Cardus called 'the most murderous onslaught I have ever known in a Test match' finished at 2.50 p.m. on the third day after five hours and twenty minutes, 376 deliveries and a century in boundaries.

The particular significance of the 254 derives, however, from Bradman's own estimation of it. While controversy attaches to other choices posthumously ascribed to him, Bradman left no room for doubt about where he ranked this feat, volunteering in *Farewell to Cricket* that it was technically the best innings of his career. 'Practically without exception every ball went where it was intended,' he opined – and 'practically' is, with Bradman, not an inconsiderable word.

This is not merely a premium endorsement either, but an insight into Bradman himself. In his restless quest for perfection, this exploit was the pinnacle of efficiency to which he himself always aspired: speed without noticeable haste, risk without obvious recklessness. If Bradman's feats now seem scarcely human, the self-scrutiny that singled this innings out implies that they cannot have been altogether unconscious.

By the same token, it is interesting that Bradman made his distinctions on a technical basis. In echoing him since, critics have been inclined to let the innings' specifications and dynamics efface its circumstances. At the time, Percy Chapman's Englishmen had Bill Woodfull's Australians very much under the cosh. The hosts held the Ashes, led 1–0 in the series and had compiled 425 in their first innings on the game's most venerable ground. The trail to a Test double century, moreover, had been blazed by only five Australians in more than fifty years of international competition.

The stage was ideally set by Woodfull and his opening partner Bill Ponsford, whose 162 for the first wicket survived every challenge save a teatime visit from King George V. Indeed, it was Woodfull whom Bradman credited with his approach: he was 'playing so finely … that I could afford to go for the bowling'.

Despite being 'naturally anxious to do well' in view of the occasion and the audience, he surged forward to meet his first ball from England's 'Farmer' White, punched it to mid-off and sauntered a single. The stroke was as clean and clear as a proclamation. 'It was,' wrote England's former captain Pelham Warner, 'as if he had already made a century.'

White, a famously parsimonious left-arm spinner, could not curb him. Nor could Maurice Tate, still probably the world's best medium-paced bowler. The young Gubby Allen and Walter Robins were harshly manhandled. Yet what was striking at once about Bradman's batting was less its power than its poise. He had held for six months the record for the biggest first-class innings: his 452 not out for New South Wales against Queensland. But this was more than humdrum accumulation of runs. It was calm, carefree, precocious; as if nobody had explained to Bradman why the occasion should daunt him and whose were the reputations he was trampling. 'Young Bradman,' said Cardus, in one of his crispest phrases, 'knocked solemnity to smithereens.'

That Cardus was present as the correspondent of *The Manchester Guardian* is history's good fortune; in cricket terms it's as if A.J.P. Taylor had been round to report the signing of Magna Carta. 'The bat sent out cracking noises; they were noises quite contemptuous,' wrote the dean of English sports journalism. 'When he batted eleven men were not enough. Lord's was too big to cover; holes were to be seen in the English field everywhere. Chapman tried his best to fill them up, but in vain.'

After tea, everyone appeared to become a spectator. To cut off Bradman's scoring seemed like trying to cap a Yellowstone geyser or a Spindletop gusher. He barely paused for the applause that greeted his 105-minute century – his third hundred in consecutive Tests – and ploughed on to the more remarkable landmark of a century in a session.

Despite Woodfull's 78-run and 170-minute head start, Bradman had caught up with his captain by the time their 160-minute partnership of 231 was ended. England's impressive total was in sight by stumps, and being judged according to an entirely different scale: suddenly no score, no statistic, no history was safe.

Given the curious queasiness that has emerged in recent years about Bradman's records, it's worth noting that the man himself knew no such taboo. When he resumed on Monday at 155, he cast intrepid and covetous eyes on the benchmark Test score of 287 by England's 'Tip' Foster. He even thought there might be something appropriate about his consigning it to oblivion: he would seize for Sydney the record set at its cricket ground twenty-seven years earlier.

With this in mind, Bradman introduced a note of care to his play before lunch, even allowing Tate to bowl him a maiden, and partner Alan Kippax to take greater initiative. Still he overhauled a double century in 245 minutes at 12.50 p.m. – becoming, at twenty-one years 307 days, considerably the youngest to achieve the feat. His lunchtime 231 was already the highest score by an Australian, the highest against England, and highest at Lord's – and still wasn't over.

Foster's citadel, in the end, did not fall. The elastic Chapman stuck his right hand aloft at extra cover to arrest a screaming drive – 'a magnificent piece of work', wrote Bradman admiringly – with the batsman 34 shy of his goal. His 254 had been made from 423 added while he'd been at the crease, and his third-wicket partnership of 192 with Kippax was another Lord's record.

Perhaps the only aspect of Bradman's innings as remarkable as the number of records is their brief durations – the cause, of course, was Bradman himself. His 254 was the Australian Test best for precisely one match; his 334 at Headingley two weeks later put everyone in the shade, including himself. Bradman's 974 runs in the five-Test series, including another 232 at the Oval in August, would remain a record seemingly beyond challenge.

It was the beginning of a sporting monopoly so unsparing it should almost have been dissolved by anti-trust regulators. To break a record is one thing; to break one's own is quite another. To make big scores is one thing; to compile them so memorably that they become associated with you forever is a mark of genuine greatness.

Inside Edge: The Greatest – Top 50 Innings by Australian Batsmen 1877–2004 (2004)

Bradman at Ninety
Sir Donald Brandname

THE AUSTRALIAN SPORTING PUBLIC is notoriously fickle, bestowing and withdrawing devotion in a blink, apt to forget even the firmest of favourites within a few years of retirement. Yet the flame of Sir Donald George Bradman, seven decades since he first made headlines, has never burned brighter.

No public appearances are expected for his ninetieth birthday on 27 August: almost a year after the passing of his beloved wife, Bradman finds them strenuous. But his continued health will be the subject of front-page encomia, and feature in evening television bulletins: an annual vigil for some years now. Whatever the tribulations of state, cricket-fancying Prime Minister John Howard will convey congratulations.

Never mind that the youngest people with clear recollection of Bradman the batsman are nudging sixty themselves, for his feats appear to be growing larger, not smaller, as they recede into antiquity. In the last decade, the cricket-industrial complex has produced a trove of books, memorabilia albums, videos, audio tapes, stamps, plates, prints and other collectables bearing the Bradman imprimatur, while the museum bearing his name at Bowral continues deriving a tidy annuity income from licensing it to coins, breakfast cereals and sporting goods.

A Bradman bat from 1930 changed hands at Philips in London last year for £21,732, a life-size Bradman bronze at Christie's in Melbourne for $65,000 six months ago. A second collecting institution has opened in his honour at Adelaide's State Library. There has been yet another reissue of Bradman's 1958 instructional bible *The Art of Cricket* and, despite full-scale biographies in 1995 and 1997, two more books are forthcoming: a compilation of tributes and a volume on Bradman's 1948 side. Sir Donald Bradman is become Sir Donald Brandname.

Mention 'the Don' in Australia and no-one mistakes it for a reference to universities or *The Godfather*. Little bits of his legend can be found everywhere. Australian state capitals boast twenty-two thoroughfares named in Bradman's honor (Victor Trumper has eight). Australians corresponding with the Australian Broadcasting Corporation do so to post office box number 9994: Bradman's totemic Test batting average, a pleasing notion of the Australian Lord Reith, Sir Charles Moses. Allan Miller, editor of *Allan's Cricket Almanack*, receives mail in Busselton at the post office box number 974: Bradman's total of runs during the 1930 Ashes series. A newspaper poll last year found that Bradman was the Australian most respondents wanted to light the flame at the 2000 Olympics: at ninety-two, it would be a feat to rank with anything he accomplished on a cricket field.

Australians can count themselves blessed that the Don is still with them. It is sixty-four years since newspapers, fearful of his prospects after a severe appendicitis, first felt the need to set obituarists on him. And Bradman was half his current age when he retired from stock-broking after a 'serious warning' from his physician. In some respects, however, Bradman himself has been supplanted in importance by Bradmyth. The idea of him is at least as important as the reality. It is odd, but not really surprising, that the best biography of Bradman was written by an Englishman: Irving Rosenwater's stupendous *Sir Donald Bradman*. And, despite the recent proliferation of Bric-a-Bradman, no-one anywhere has tangibly added to the sum of human knowledge about the Don in twenty years. The most recent Bradman biography, Lord Williams' *Bradman: An Australian Hero*, is a case in point: of 428 footnotes, 244 referred to four titles, two of them previous Bradman biographies.

At one time, it was Bradman who sought Garbo-like quietude, no less than he deserved after more than four decades seldom out of the public eye. Nowadays, though, Australians do just as much to preserve that distance. The last locally produced Bradman biography

– Roland Perry's *The Don* (1995) – had as much substance as a comic strip. The last public interview with Bradman – two hours broadcast in May 1996 by Channel Nine's top-rating current affairs host Ray Martin on the basis of a corporate donation to the Bradman Museum – was what *Private Eye* used to describe as a journey to the province of Arslikhan.

It may justly be asked what more of the Bradman saga begs understanding. After all, the Greatest Story Ever Bowled To is so beguiling as it is: uncoached boy from the bush rises on merit, plays for honour and glory, puts Poms to flight, becomes an intimate of sovereigns and statesmen, retires Cincinnatus-like to his unostentatious suburban home. But turning Bradman into Mr 99.94 is a bit like reducing Einstein to Mr $E=mc^2$. Read most Bradmanarama and you'd be forgiven for thinking that his eighty Test innings were the sum of him. His family is invisible. Precious little exists about Bradman's three decades as an administrator. There is next to nothing about his extensive business career. And no-one, I think, has grasped what is perhaps most extraordinary about Bradman: his singularity as a man as well as a cricketer. For the great irony of his beatification is that he was never, as one might imagine, an acme of Australianness.

For most Australian boys, for instance, participation in sport is a rite of passage, an important aspect of socialisation. Yet, if Bradman developed close cricketing pals in his Bowral boyhood, they kept remarkably schtum afterwards. The rudimentary game with paling bat and kerosene tin wicket in some urban thoroughfare is one of Australian cricket's cosiest images: think of Ray Lindwall and his cobbers playing in Hurstville's Hudson Street, trying to catch the eye of Bill O'Reilly as that canny old soul walked by; or of the brothers Harvey playing their fraternal Tests behind the family's Argyle Street terrace in Fitzroy. Bradman's contribution to the lore of juvenile cricket, by contrast, is one of solitary autodidacticism, his water-tank training ritual with golf ball and stump.

That carapace hardened as Bradman reached cricketing maturity, and set him still further apart. Where the archetypal Australian male is hearty and sanguine, priding himself on good fellowship, hospitality and ability to hold his alcohol, Bradman was private, reserved, fragile of physique and teetotal. Where the traditional Australian work ethic has been to do just enough to get by, Bradman was a virtuoso who set his own standards and allowed nothing to impede their attainment.

Australia in the late 1920s, moreover, was not a country that seemed likely to foster an abundance of remarkable men. It was a small subsidiary of Empire, with an ethnically and culturally homogenous population of six million. Even that big bridge was still to come. There were extremes of wealth and poverty, but social mobility was restricted both by economic hardship and prevailing belief in an underlying social equality. Visiting Australia for the first time in the year of Bradman's first-class debut, American critic Hartley Grattan was amazed by the vehemence of this latter faith: 'Australia is perhaps the last stronghold of egalitarian democracy ... The aggressive insistence on the worth and unique importance of the common man seems to me to be one of the fundamental Australian characteristics.' As D.H. Lawrence described it in his novel of 1920s Australia, *Kangaroo*: 'Each individual seems to feel himself pledged to put himself aside, to keep himself at least half out of count. The whole geniality is based on a sort of code of "You put yourself aside, and I'll put myself aside." This is done with a watchful will: a sort of duel.'

Bradman, however, was not a 'common man', and he assuredly did not 'put himself aside'. R.C. Robertson-Glasgow recalled that, at his first meeting with Bradman at Folkestone in September 1930, the Australian was surrounded by piles of correspondence to which he was steadily reaming off replies. 'He had made his name at cricket,' wrote Crusoe. 'And now, quiet and calculating, he was, he told me, trying to capitalise his success.'

The times may have been ripe for such individual aspiration. Certainly, Bradman's benefactors on that tour had no difficulty singling him out for gifts and gratuities, not least the Fleming and Whitelaw soap magnate Arthur Whitelaw who spontaneously bestowed £1000 on Bradman after his Headingley 334. But nothing before or since has paralleled the Caesar-like triumph that Bradman's employer Mick Simmons Ltd organised for him when the team returned to Australia, where he travelled independently of his team and was plied with public subscriptions and prizes in Perth, Adelaide, Melbourne, Goulburn and Sydney.

It was the beginning of a career in which Bradman showed conspicuous aptitude for parlaying his athletic talent into commercial reward. Leaving Mick Simmons Ltd in 1931, he signed a tripartite contract worth more than £1000 a year with radio station 2UE, retailer F.J. Palmer and Associated Newspapers (proprietor Robert Clyde Packer, grandfather of Kerry). He spruiked bats (Wm Sykes), boots (McKeown) and books (three while he was playing, two afterwards), irked the Australian Board of Control for International Cricket by writing about cricket in apparent defiance of its dictates, deliberated over effectively quitting Test cricket to accept the Lancashire League shilling for the 1932 season, swapped states in 1935 to further his career. At a time when Australian industry lurked behind perhaps the highest tariff barriers on earth, Bradman was an ardent apostle of the free market.

No dispute that Bradman deserved every penny and more. No question of undue rapacity either. As that felicitous phrase-maker Ray Robinson once expressed it, the Don did not so much chase money as overhaul it. Equally, however, Bradman's approach betokens an elitism uncharacteristic of Australia at the time and not a quality many today would willingly volunteer as a national hallmark.

It was this impregnable self-estimation – not arrogance, but a remarkable awareness of his entitlements – that distanced Bradman from his peers. Some criticisms of the Don by playing contemporaries were undoubtedly actuated by jealousy. All the same, he seems to have been incapable of the sort of gesture that might have put comrades at their ease. In his autobiography *Farewell to Cricket*, he commented: 'I was often accused of being unsociable because at the end of the day I did not think it my duty to breast the bar and engage in a beer-drinking contest.' It is a curious perspective on cricket's social conventions, with the implication, as John Arlott put it, that Bradman had 'missed something of cricket that less gifted and less memorable men have gained'.

Bradman's playing philosophy – that cricket should not be a career, and that those good enough could profit from other avenues – also seems to have borne on his approach to administration. Biographers have disserved Bradman in glossing over his years in officialdom. His strength and scruples over more than three decades were exemplary; the foremost master of the game became its staunchest servant. But he largely missed the secular shift towards the professionalisation of sport in the late '60s and early '70s, which finally found expression in Kerry Packer's World Series Cricket.

Discussing the rise of World Series Cricket, Bradman told Williams in January 1995 he 'accepted that cricket had to become professional'. Yet, as Dr Bob Stewart comments in his recent work on the commercial and cultural development of post-war Australian cricket *I Heard It on the Radio, I Saw It on the Television,* cricket wages declined markedly in real terms during the period that Bradman was Australian cricket's *eminence gris.* When Bradman retired, the home Test fee was seven times the average weekly wage. A quarter of a century later it was twice the average weekly wage. Ian Chappell contended in *The Cutting Edge* that the pervasion of Bradman's attitude to player pay within the Australian Cricket Board 'contributed to the success World Series Cricket officials had when a couple of years later they approached Australian players with a contract'.

Perhaps these paradoxes of the Bradman myth relate something about the complex Australian attitude to sport. As the Australian social commentator Donald Horne once put it: 'It is only in sport that many Australians express those approaches to life that are un-Australian if expressed any other way.' But, as Bradman enters his tenth decade fit for both commodification and canonisation, two questions seem worth asking, with apologies to C.L.R. James.

First: What do they know of Bradman who only cricket know? Surely it's possible in writing about someone who has lived for ninety years to do something more than prattle on endlessly about the fifteen or so of them he spent in flannels – re-circulating the same stories, the same banal and blinkered visions – and bring some new perspectives and insights.

Second: What do they know of cricket who only Bradman know? A generation has now grown up in Australia that regards cricket history as 6996 and all that. Where are the home-grown biographies of Charlie Macartney, Warwick Armstrong, Bill Woodfull, Bill Ponsford, Lindsay Hassett, Keith Miller, Neil Harvey, Alan Davidson, Richie Benaud, Bob Simpson, even Ian Chappell and Dennis Lillee, plus sundry others one could name? Such is the lava flow from the Bradman volcano, they are unlikely to see daylight.

So enough with the obeisances already. Yes, Bradman at ninety is a legend worth saluting. But as the American journalist Walter Lippman once said: 'When all think alike, none are thinking.'

Wisden Cricket Monthly August 1998

Bradman Redux
The Greatest No-longer-living Australian?

THE STORY OF SIR DONALD BRADMAN always involved more than cricket – more even than sport. One can only marvel at the statistics, and the only batting average that nobody need ever look up. Yet, in its degree and duration, especially in Australia, Bradman's renown is as much a source of wonder: he was, as Pope wrote of Cromwell, 'damned to everlasting fame'.

Few figures, sporting or otherwise, have remained an object of reverence for more than half a century after the deeds forming the basis of their reputation. Fewer still can have justified an autobiography at the age of twenty-one, and the last of many biographies at the age of eighty-seven, without surfeiting or even satisfying public curiosity. But with Bradman's death on 25 February 2001 comes a question: what difference will it make to his legend that, for the first time, it has obtained a life independent of his corporeal existence?

On the face of it, the answer appears: not much. In Australia, the Bradman story detached from the Bradman reality some time ago, taking on the qualities of myth: as defined by Georges Sorel, a faith independent of fact or fiction, where the normal processes of attestation and falsification are suspended by unconscious agreement. In the 1990s, in particular, a sort of cultural elision turned Bradman from great batsman to great man: by Prime Minister John Howard's lights 'the greatest living Australian', elevated not merely by his feats but by 'the quality of the man himself'. Inspired by Bradman, Howard claimed, Australians could make 'any dream come true', and 'not just in cricket but in life'.

This was a new experience for Australians. Poet and critic Max

Harris once declared that 'the Australian world is peopled by good blokes and bastards, but not heroes'. Yet, by the end of his life, Bradman was not merely a 'hero'; he had also been, in effect, retrofitted as a 'good bloke' – that peculiarly Australian formulation implying, essentially, someone like everybody else. The actuality is, of course, that Bradman was an individual to whom few beyond his immediate family circle were close, whose private life was sedulously guarded, and whose inner-most convictions went mostly unaired; his 'good blokeness' is, accord-ingly, more or less unascertainable. But, as Sorel observed: 'People who are living in this world of myths are secure from all refutation.'

Why this happened can only be conjectured. Perhaps it was the self-image as a sports-loving nation to which Australians cling with such tenacity. Perhaps it was part of the recrudescence of Australian conservatism (it's hard to imagine Howard's republican predecessor Paul Keating singling Bradman out as the 'greatest living Australian'; indeed, perhaps that partly accounts for his electoral failure). It may be as simple as Bradman keeping his head while all about him were losing theirs. He avoided the public eye, courted no controversies, and scorned fame's usual accoutrements. It would be an exaggeration to describe him as an outstanding citizen, at least as this is commonly understood: he was not a conspicuous philanthropist or supporter of causes, and the duty to which he devoted himself most diligently was simply being Donald Bradman, tirelessly tending his mountainous correspondence and signing perhaps more autographs than anyone in history. But there is no doubt that, standing aloof from modern celebrity culture, he retained a dignity and stature that most public figures sacrifice.

This being the case, it may be that the 'greatest living Australian' segues smoothly into becoming the greatest dead one: Bradman certainly won't be doing anything controversial now. But there are reasons he may not: countries progress, palates jade, reputations are reassessed. There is already something anomalous about Bradman's standing in Australia. National heredity and heritage during Bradman's playing career were almost exclusively British; as Dr Greg Manning has noted, Australia's cultural homogeneity in the 1930s and 1940s was a precondition of the Bradman phenomenon. But since the late 1980s, British and old Australian components have accounted for less than half of Australia's population. Even the idea of a monolithic Australian mourning of Bradman's death was mostly a media assumption. *The Chaser*, a satirical newspaper, expressed this drolly with a story headlined: 'Woman unmoved by Bradman's death'. This fantasised of a forty-two-year-old woman who 'surprised reporters when she was unable to recite Bradman's famous batting average or other miscellaneous statistics', who 'denied suggestions the cricketer had been an inspiring influence on her life, and said she had no regrets whatsoever that she never saw him play'.

A significant role in the ongoing Bradman story will be played by the Bradman Museum at the cricketer's boyhood home of Bowral in New South Wales' southern highlands. So far, the museum is a success story, as unique as its eponymous inspiration: a collecting institution for cricket that is largely self-funding, thanks partly to royalties from the Bradman name worth between a quarter and half a million dollars annually. And for this, Bradman himself can take a deal of credit. It was he, says the museum's thoughtful curator Richard Mulvaney, who 'recognised that the museum would always have problems getting enough money to do the things it wanted to do', and ten years ago vouchsafed the commercial rights to his name and image. 'We wouldn't have dared ask,' says Mulvaney.

The museum's future financial security was consolidated in August 2000 by an unforeseen consequence of the South Australian Government's decision to re-name an Adelaide thoroughfare in honour of its favourite adopted son. When several businesses in the new Sir Donald Bradman Drive proposed exploiting their location in advertising, the Bradman family sought government intervention. One enterprise, in particular, set heaven in a rage. 'When the sex shop registered as Erotica-on-Bradman,' Mulvaney recalls, 'we couldn't have had a better example of the ways the name could be misused.' The subsequent amendment to Australia's Corporations Law – preventing businesses from registering names that suggested a connection with Bradman where it did not exist – invested the Don with a commercial status previously the preserve of members of the royal family. But there was some sense to it, and the amendment has already proved its worth in the relative scarcity of necrodreck in Australia after Bradman's death. In the longer term, too, it should secure for cricket the financial fruits of the Don's legacy, for which the game can be grateful.

Enhanced power over the Bradman franchise sets the museum challenges. Its charter is to be 'a museum of Australian cricket commemorating Sir Donald Bradman'. The legislation, explicit acknowledgement that Bradman's name is a commodity with commercial value, subtly firms the emphasis on the latter role – something the museum recently recognised by signing an agency agreement for the exploitation of the name with Elite Sports Properties, a six-year-old sports management group now owned by London-based Sportsworld Media plc.

To be fair, Mulvaney perceives the dilemma. He is adamant that the museum will not, in his watch, become a one-man hall of fame: 'While it's in my interest to continue putting Bradman's name forward, we shouldn't take his name out of context. We have to be careful that we don't allow excesses where the name takes on a completely different meaning.' The responsibilities inherent in ESP's role also weigh heavily on its Andrew Stevens. 'We feel very honoured to be representing the greatest Australian icon who ever lived,' he says. 'And protecting it, which is quite daunting. It's very precious. In fact, if someone strikes out at it, or is reckless with it, you tend to feel quite aggrieved.'

To how much protection, though, is Bradman entitled? And what constitutes 'recklessness' where the name is concerned? These questions were raised last year in an episode involving two letters by Bradman to Greg Chappell dating from the World Series Cricket schism in 1977. After defecting to Kerry Packer's enterprise, Chappell found himself *persona non grata* in his adopted state of Queensland, and confided in a *Brisbane Telegraph* journalist some disparagements of the quality of local cricket administrators that Bradman had made to him some years earlier. When these were published, however, Chappell received a letter of rebuke for what Bradman viewed as a breach of trust; an exchange of correspondence followed in which Bradman enlarged on certain personal philosophies, as well as lamenting the frequent press misrepresentation to which he had been subject.

These were important artefacts. They evoked both the passion of the moment – for this was a time when tempers in cricket were frayed on all sides – and the personality of the writer. In some respects, it is from such flashes of candour that biographies are built; thus Plutarch's famous remark in his life of Alexander the Great that 'very often an action of small note, a short saying, or jest, will distinguish a person's real character more than the greatest sieges or the most important battles'. But when the letters were auctioned by Christie's in July 2001, Chappell was roundly censured for another breach of trust, that of Bradman's privacy. An anonymous consortium of businessmen styling themselves SAVE – Some Australians Value Ethics – intervened to purchase the letters, plus three others, for $25,870. They were presented to Mulvaney at the MCG on what would have been the Don's ninety-third birthday.

Even ignoring the question of whether it's possible to breach the privacy of the dead, it was a curious interlude. It went unremarked, for example, that the letters had been quoted from before, without protest, in Adrian McGregor's excellent 1985 Chappell biography. At best, SAVE's intercession was a well-meaning attempt to avoid media sensationalising, which perversely made the letters' contents seem more sensational than they were; at worst, it evinced a disconcerting cultural timidity in Australia, a squeamishness about anything other than the 'approved' version of the Bradman story. In an interview with *Sporting Collector*, Mulvaney was quoted as saying that the Bradman Museum 'considered the contents not necessarily to be in the public interest' and that it had 'no option but to keep them private' – which made it sound like an issue of national security.

In an interview with me, Mulvaney said that the magazine had misrepresented his views: his concern was purely with copyright, which he felt Christie's had breached by reproducing the letters in its catalogue, and insisted that the museum had no wish to inhibit study of Bradman material. When I sought permission to quote the letters, however, Sir Donald's son John withheld it; he stated, albeit politely, that this was his present policy on all such inquiries.

One can sympathise with John Bradman's position. No policy will satisfy everyone. Permission for all will invite excess and exploitation; permission for some will smack of favouritism; permission for none will be construed as censorship. Yet attempting to control what is published about Bradman in such a way will carry risks. Revisionists will move in regardless; indeed, they already have. In November 2001, *The Australian* published two articles concerning 'the Don We Never Knew', the work of a foot-slogging journalist, David Nason. One revealed what appeared vestiges of a family rift between the Don and his Bowral kin, disclosing the view of Bradman's nephew that 'Don Bradman only ever cared about Don Bradman'; the second examined how the alacrity with which Bradman took over the defunct stock-broking firm of the disgraced Harry Hodgetts had antagonised Adelaide's establishment.

Much the same criticisms of invasiveness were advanced as in the Chappell case. *Sunday Telegraph* columnist Mike Gibson scorned the first article as 'pathetic', 'despicable', 'bitter and twisted'; the second he dismissed as *lese majeste*: 'You know, I couldn't be bothered reading it. First, his family. Then, his finances. I couldn't stomach any more smears against someone who cannot answer back because he's dead.' Yet what was perhaps truly noteworthy about the articles was not that they appeared, but that nothing resembling them had appeared before, that none of Bradman's *soi-disant* 'biographers' had treated his family and financial lives other than perfunctorily. Indeed, it may be that Bradman's most ardent apologists end up doing him the gravest disservice. Identifying a great sportsman isn't difficult; the criteria are relatively simple. Designating a great man entails rather more than a decree, even prime ministerial; some intellectual and historic contestation must be involved. Credulity invites scepticism – and, unlike trade names, reputations cannot be declared off limits by the wave of a legislative wand.

Wisden Cricketers' Almanack 2002

Eddie Gilbert and Mark Waugh
Dark Victory

BROWSERS IN THE BOOKSHOP SHELVES allocated to sport this summer will probably find *Eddie Gilbert* by Mike Colman and Ken Edwards and *Mark Waugh* by James Knight snuggled quite close together. Both, after all, are biographies of Australian cricketers, written by journalists, and published by firms with strong sporting backlists. But their proximity will be misleading. Cricket contains few less similar careers, and can have generated few more different styles of story: indeed, reading them consecutively is to appreciate how stealthily our understanding of 'biography' has been elasticised.

Eddie Gilbert should be far better known. An Aboriginal raised on a controlled settlement at Barambah in Queensland, his speed as a fast bowler unsettled and unseated the mighty Bradman in a celebrated over at the Gabba seventy years ago. Having run the racist gauntlet of the time, though, Gilbert was tripped up by alcoholism and womanising, and prostrated by dementia that confined him to an institution for the last twenty years of his life. It is a story rich in narrative potential – the inspiration, in fact, behind David Forrest's enchanting short story 'That Barambah Mob' (1965). But what Gilbert has needed for many years has been a version of his story shorn of drama, ornament and effect, assembling the known facts and blending them with new. The new collaboration between Mike Colman and Ken Edwards fills this need admirably.

The book really belongs to Edwards, a lecturer at Queensland University of Technology who spent seven years researching Gilbert's story for a Ph.D. Grasping an idea that some academics seem to find elusive, that a thesis is no more a book than a lump of wood is a cricket bat, he then had the work rewritten in an accessible but analytical style by

Courier-Mail columnist Colman. The result is a distinct Australian rarity: a book about sport that is both readable and rigorous. Edwards has had access to settlement records and correspondence, interviewed widely among Gilbert's indigenous contemporaries, former team-mates and rivals, and flinched from nothing.

Gilbert was a freakish cricketer, handicapped by a lack of stamina, and eventually tainted by suspicions about the legality of his action, but probably faster than any Australian bowler at the time, and at least in his home state a popular hero. Not surprisingly, his story has been encrusted by legend and lore, and there's much to admire about *Eddie Gilbert*'s methodical approach to distinguishing fact from fancy. Old stories that have been 'authenticated' by repetition, such as Gilbert's living in a tent in the garden of a Queensland Cricket Association official, are proven false. Edwards' research would have been improved only by a comparison between responses to Gilbert and representations of other black athletes of the period, which might have revealed some illuminating inter-texts: the stories of Gilbert's bowling barefoot which Edwards concludes are untrue, for example, seem to echo the tale of 'Float' Woods pleading with his West Indian captain in 1900 to let him 'feel de pitch wid de toe'.

The weaknesses of *Eddie Gilbert* are to be found, on occasion, in the prose. There are some contestable judgements ('To the majority of Australians in the early 1900s, sport was seen as fun, a diverting pastime') and some avoidable clichés ('His word was law and it was a law he policed with an iron fist'). Then, at perhaps the pivotal point, Gilbert's defeat of the Don for a duck in November 1931, the narrative loses shape and form, being reported as a kind of imagined re-creation straight out of Creative Writing 101. These are, though, relatively rare lapses: for the most part, Colman does his wordsmithing with unobtrusive competence.

On one level, *Eddie Gilbert* will not satisfy, and probably couldn't. A quiet man who left little impression on contemporaries save his speed, and whose recorded remarks were restricted to a few corner-of-the-mouth confidences to journalists, he remains an elusive subject. This, though, matters less than it might: the reactions to Gilbert are probably as instructive as anything he might have said or thought. Cooperation of a subject, furthermore, does not guarantee note-worthy biography, as may be deduced from James Knight's *Mark Waugh: The Biography*.

This is one of those 'authorised biographies' that now abound in sport, and are increasingly popular among other demi-celebs. They are, essentially, slightly modified autobiographies, in which the ghost writer hardens from ectoplasm into slightly more palpable form as a narrator, while the subject's views are interposed as quotations rather than revealed in the first person. With further quotations from others tactfully scattered throughout the text, this suggests a conventional arms-length biographical portrait; in fact, the views are carefully sanitised so as to show the subject in the best possible light.

There's no cause for alarm. Buying memoirs by a modern sportsman is like making a donation to their testimonial fund and being given a free book as a receipt, not so much to read, but to symbolise your allegiance. Even by genre standards, however, *Mark Waugh* is remarkably free of charm. Knight is a mediocre writer ('Young. It was the most fitting way to describe Mark before his first Sheffield Shield match'); worse, he's a mediocre writer with a flair for the ornate ('Nowadays, one-day matches are like confetti – the brilliance they offer for a fleeting moment or two generally becomes nothing more than scraps flicked from the memories of players'). His attempts to describe places are buttock-clenchingly bad: 'There is a touch of "The Man From Ironbark" about Jalandhar, for in this Sikh town "flowing beards are all the go". So too the brilliantly coloured turbans that are matched only by the dazzling smiles of the people.'

The world according to Knight, in fact, divides roughly into those places that are poor and happy like Sri Lanka ('Despite the sadness, communities continue to smile with a warmth that hovers well above the temperature in this equatorial and political hotspot') or the West Indies ('The region is poor, yet the people are rich in character and warmth'), and those that are poor and bearded like Pakistan ('Men with flowing beards kneel in prayer. A bus belches filthy black smoke … This is Pakistan, the toughest country for an Australian cricketer to tour'). Not that Mark Waugh's travels are much influenced by either smiles or beards: when on tour, we learn, he seldom leaves hotels, takes no photos, and finds the foreigners, well, a bit on the whiffy side; 'it's a bit off-putting having a bloke standing next to you with a gun, but … actually their body odour is more unsettling'. Fortunately, Third World folk generally know their place: 'We know we're in a privileged position when we go there because people just want to worship us and treat us like kings.'

Not all of the 380 pages here can be cribbed from the postcards on Knight's fridge, so readers must grind their way through Waugh's career innings by innings. The effect is repetitive at best; inane at worst. It is actually a curious sensation to read *Mark Waugh* immediately after *Eddie Gilbert* – the effect of passing from a subject who talked so little yet said so much, to one who talks so much and says so little.

Australian Book Review September 2002

Richie Benaud
The Face of Cricket

'DID YOU EVER PLAY CRICKET FOR AUSTRALIA, Mr Benaud?' In his *On Reflection*, Richie Benaud recalls this humbling question from a 'fair-haired, angelic little lad of about twelve', one of a group of six autograph seekers who accosted him at the Sydney Cricket Ground 'one December evening in 1982'.

'Now what do you do?' Benaud writes. 'Cry or laugh? I did neither but merely said yes, I had played up to 1963, which was going to be well before he was born. "Oh," he said. "That's great. I thought you were just a television commentator on cricket."' Autograph in hand, the boy 'scampered away with a "thank you" thrown over his shoulder'.

It's a familiar anecdotal scenario: past player confronted by dwindling renown. But the Benaud version is very Benaudesque. There is the amused self-mockery, the precise observation, the authenticating detail: he offers a date, the number of boys and the appearance of his interlocutor, whose age is cautiously approximated.

In his story, Benaud indulges the boy's solecism, realising that it arises not merely from youthful innocence, but from the fact that 'he had never seen me in cricket gear, and knew me only as the man who did the cricket on Channel Nine'. Then he segues into several pages of discussion of the changed nature of the cricket audience, ending with a subtle self-disclosure: 'Some would say such a question of that kind showed lack of respect or knowledge. Not a bit of it … What it did was show an inquiring mind and I'm all in favour of inquiring minds among our young sportsmen. Perhaps that is because I had an inquiring mind when I came into first-class cricket but was not necessarily allowed to exercise it in the same way as young players are now.'

I like this passage; droll, reasoned and thoughtful, it tells us much about cricket's most admired and pervasive post-war personality. It is the voice, as Greg Manning phrased it in last year's *Wisden Australia*, of commentary's 'wise old king'. It betrays, too, the difficulty in assessing him: in some respects, Benaud's abiding ubiquity in England and Australia inhibits appreciation of the totality of his achievements.

In fact, Benaud would rank among Test cricket's elite leg spinners and captains even if he had never uttered or written a word about the game. His apprenticeship was lengthy – thanks partly to the prolongation of Ian Johnson's career by his tenure as Australian captain – and his first twenty-seven Tests encompassed only 73 wickets at 28.9 and 868 runs at 20.68. Then, as Johnnie Moyes put it, came seniority and skipperhood: 'Often in life and in cricket we see the man who has true substance in him burst forth into stardom when his walk-on part is changed for one demanding personality and a degree of leadership. I believe that this is what happened to Benaud.' At the peak of proficiency he attained in his next twenty-three Tests, until shoulder injury in May 1961 impaired his effectiveness, his record was 131 wickets at 22.7 and 830 runs at 29.

Australia, moreover, did not lose a series under Benaud's leadership, although he was defined by his deportment as much as his deeds. Usually bareheaded, and with shirt open as wide as propriety permitted, he was a colourful, communicative antidote to an austere, tight-lipped era. Jack Fingleton likened Benaud to Jean Borotra, 'the Bounding Basque of Biarritz' over whom tennis audiences had swooned in the 1920s; *Wisden* settled for describing him as 'the most popular captain of any overseas team to come to Great Britain'.

One legacy of Benaud's is the demonstrative celebration of wickets and catches, a conspicuous aspect of his teams' communal spirit, today *de rigueur*; another is a string of astute, analytical books, including *The Way of Cricket* (1960) and *A Tale of Two Tests* (1962), among the best by a cricketer during his career. 'In public relations to benefit the game,' Ray Robinson decided, 'Benaud was so far ahead of predecessors that race-glasses would have been needed to see who was at the head of the others.'

Benaud's reputation as a gambling captain has probably been overstated. On the contrary, he was fastidious in his planning, endlessly solicitous of his players, and inclusive in his decision-making. Benaud receives less credit than he deserves for intuiting that 'eleven heads are better than one' where captaincy is concerned; what is a commonplace now was not so in his time. In some respects his management model paralleled the 'human relations school' in organisational psychology inspired by Douglas McGregor's *The Human Side of Enterprise* (1960). Certainly, Benaud's belief that 'cricketers are intelligent people and must be treated as such', and in 'an elastic but realistic sense of self-discipline' could be transliterations of McGregor to a sporting context.

Ian Meckiff defined Benaud as 'a professional in an amateur era', a succinct formulation which may partly explain the ease with which he has assimilated the professional present. For if a quality distinguishes his commentary, it is that he calls the game he is watching, not one he once watched or played in. Most past players turned pundits succumb at some stage to *laudator temporis acti*. Benaud never has. When Simon Katich was awarded his baggy green at Headingley last year, it was Benaud whom Steve Waugh invited to undertake the duty.

Benaud's progressive attitude to cricket's commercialisation – sponsorship, television, the one-day game – may also spring partly from his upbringing. In *On Reflection*, he cites the story of his father Lou, a gifted leg spinner whose cricket ambitions were curtailed by his posting to the country as a schoolteacher for twelve years. Benaud describes two vows that his father took: 'If ... there were any sons in his family he would make sure they had a chance [to make a cricket career] and there would be no more schoolteachers in the Benaud family.'

At an early stage of his first-class career, too, Benaud lost his job with an accounting firm that 'couldn't afford to pay the six pounds a week which would have been my due'. He criticised the poor rewards for the cricketers of his time, claiming that they were 'not substantial enough' and that 'some players ... made nothing out of tours', while contending rather adventurously that 'cricket is now a business' as far back as 1960.

Those views obtained active expression when he aligned with World Series Cricket twenty-five years ago – it 'ran alongside my ideas about Australian cricketers currently being paid far too little and having virtually no input into the game in Australia' – and Benaud's contribution to Kerry Packer's venture as consultant and commentator was inestimable: to the organisation he brought cricket know-how, to the product he applied a patina of respectability. Changes were wrought in cricket over two years that would have taken decades under the game's existing institutions, and Benaud was essentially their front man.

In lending Packer his reputation, Benaud ended up serving his own. John Arlott has been garlanded as the 'voice of cricket'; Benaud is indisputably the 'face of cricket', in both hemispheres, on both public broadcasters and commercial networks, over generations. If one was to be critical, it may be that Benaud has been too much the apologist for modern cricket, too much the Dr Pangloss: it is, after all, difficult to act as impartial critic of the entertainment package one is involved in selling. Yet Benaud is one of very few certifiably unique individuals in cricket history. From time to time, one hears mooted 'the next Benaud'; one also knows that there cannot be one.

Wisden Cricket Monthly January 2002

Sir Garfield Sobers
The Awkward Hero

IN *Cricket in the Leagues*, JOHN KAY describes a match in which Sir Garfield Sobers represented Norton, and was hit back over his head. Sobers reportedly turned, ran towards the boundary shouting 'leave it to me', and took the catch just in front of the sightscreen.

'Leave it to me' – and you could. To Ray Robinson, Sobers was 'evolution's ultimate specimen in cricketers'. He combined faculties of the game as none other, holding down at some stage every place in the West Indian order, bowling every variety of left-arm delivery, fielding in every position with uniform excellence, and captaining with a knowledge that his predecessor Sir Frank Worrell described as encompassing 'everything' – this for two decades, with only the barest attenuation in performance as age encroached. Incorporate Sobers' matches leading World XIs in England and Australia, and his international record stretches from horizon to horizon: almost 9000 runs at 58, 265 wickets at 33, 121 catches.

All this was accomplished, furthermore, with a unique elasticity and *elan*. Of his batting, Barry Richards once observed that Sobers was 'the only 360-degree player in the game'; that is, his follow-through ended where his pick-up began, swinging 'right through every degree on the compass'. The generosity of swing was matched by a generosity of spirit, for Sobers was also a 'walker' throughout his career: not only a mark of sportsmanship, but indicative of an easeful confidence in his abilities that needed no help from luck. And one notes of his bowling that, despite operating under front- and back-foot laws, he bowled only a handful of no-balls in his entire career; such was his instinctive calibration for cricket.

In the post-colonial West Indies, not surprisingly, Sobers was a point of pride, almost a divine. C.L.R. James – and none wrote better of Sobers than James – saw in him 'the living embodiment of centuries of tortured history', and 'a West Indian cricketer, not merely a cricketer from the West Indies'. The appointment of Worrell to lead the West Indies to Australia in 1960–61 is generally regarded as the defining moment in the chronicles of Caribbean cricket. But as Brian Stoddart suggests in *Liberation Cricket: West Indies Cricket Culture*, Sobers' succession in June 1964 was 'in many ways ... the real break with the past'. Worrell was courtly, tactful, educated at an English university, a Freemason; Sobers' background was both indigenous and indigent. He was raised by a widow in a small shack in Bridgetown's Bay Land, a tenantry created for plantation workers after the cessation of slavery in 1838, between docks that were a crucible of industrial disputation and a local cricket club that was for whites only. His rise from exclusion to eminence, Hilary Beckles contends in *The Development of West Indies Cricket*, sent 'a signal of hope for collective redemption'.

In some senses, however, Sobers was an awkward hero in the cause of West Indian self-affirmation through sport. He did not view Test match defeat as setting back the struggle for national prestige: when he scandalised countrymen with his costly declaration in the Trinidad Test of March 1968, he could only reply that 'cricketers are entertainers'. He was awkwardly apolitical: he explained his controversial visit to Rhodesia for a double-wicket competition in September 1970 by saying he 'knew nothing of politics at the time', while his avowal that 'there should be no barriers in sport' smacked of apology for sporting contact with South Africa (Sobers, in fact, makes clear in *The Changing Face of Cricket* that he did not himself play in South Africa only because he had 'had my fill of publicity and criticism' during the Rhodesian affair, and 'did not want to put up with all that again'; it had 'nothing to do with what was happening in the country'). He was as slow to militancy as some of his successors were quick. It is interesting, for example, to contrast Sobers' and Viv Richards' responses to Tony Greig's vow to make the 1976 West Indians 'grovel'. In *Hitting Across the Line*, Richards recalls it as a 'racist dig' that 'gave strength to a lot of the West Indian guys'. In *Twenty Years at the Top*, Sobers dismisses it as a throwaway line from a 'talking captain', observes that West Indian players were 'upset' without commenting why, then adds that Greig probably 'relished the extra challenge his interview provoked'.

On the face of it, then, there is a tension here. Sobers was, for many of his countrymen, a cultural embodiment – in C.L.R. James' words, 'the fine fruit of a great tradition'. Yet from dimensions of that culture, he could appear remote, removed. Why? It may be instructive to consider one of the most intriguing events in West Indian cricket history – or, to be precise, non-events, for it did not happen.

After his first few years in Test cricket, Sobers followed the well-trodden route of many West Indian cricketers into the English leagues, representing Radcliffe in the Central Lancashire League between 1958 and 1962. After his feats in Australia in 1960–61, he was also lured to South Australia for the ensuing three seasons, sponsored by Prospect Cricket Club and *The Adelaide Advertiser*, where he inspired their first Sheffield Shield win in a decade. 'My kind of cricket is world cricket,' he explained breezily in *Cricket Crusader*. 'And the jet plane makes all-year-round cricket possible for the individual world player.'

It was as an 'individual world player' representing South Australia that, in early 1963, the invitation from the West Indies Cricket Board of Control to tour England later that year found him. The tour fee, a derisory £800, irked Sobers: he could earn far more in a summer of one-day-a-week league cricket. He equivocated (as did Wes Hall, then representing Queensland). He discussed the tour with both Sir Donald Bradman and Richie Benaud. Both urged him to go, although Bradman agreed that 'with your standing, you should get more money than anyone in the game', and Sobers remained ambivalent. What finally persuaded him was a 'strongly worded' letter from Frank Worrell, reminding Sobers of his patriotic duty.

Many sportsmen of the period, of course, were struggling to reconcile the prestige of international sport with its paltry rewards. But Sobers' position was more complex than, say, an Australian tennis player choosing between the patriotic tug of Davis Cup and the enticements of the professional circuit. Cricket was at the heart of West Indian cultural identity, more than a game, more than a job: its devotees were unready for an 'individual world player' who decoupled sport and politics.

None of which is to say that Sobers should have rejected Worrell's overtures; one rejoices that he didn't, for his 322 runs and 20 wickets were decisive in one of Test cricket's greatest series. Nor is it to say that Sobers went altogether unrewarded. When England admitted overseas professionals in 1968, Sobers was comfortably the best-paid, on £5000 a season; he says himself in *Changing Face* that he 'lived well in my twenty years in the game', even if 'my mind boggles at the earnings of today's players' when compared with his own. But there was a paradox in his position: in participating in the emancipation of West Indian cricket, the outstanding cricketer of his generation had to consent to his own continued economic exploitation. To optimise his rewards, his only recourse was to play virtually without interruption throughout his career: in Test, first-class, league and one-day cricket, he batted almost 900 times for nearly 40,000 runs, and bowled more than 92,000 deliveries for close to 2000 wickets.

Sobers' experiences account for his rallying to the cause of Kerry Packer's World Series Cricket in 1977, 'because I wanted to see the leading players earn more money from the game'. He oversaw the toss in the First Supertest, lent his name to the Sir Garfield Sobers Trophy for which the Australian, West Indian and World XIs competed, and acted as a Packer ambassador in Australia and the West Indies. It also explains his reluctance to criticise the West Indian rebel tours of South Africa in 1983–85. While West Indian cricketers of a more recent generation like Richards, Joel Garner and Michael Holding were outspoken in their condemnations, Sobers felt a simpatico with players 'whose futures were far from secure and who saw an opportunity to put themselves on a more financially viable footing', and thought that they 'had been unfairly treated'. In what you might call the Jamesian chain of cricket heroes – linking Constantine, Headley, Worrell, Sobers and Richards – Sobers' is an unusual presence, one who did all things on the cricket field with ease, but who encountered the difficulty off it of having to be all things to all men.

Wisden Cricket Monthly February 2002

Wasim Akram
Reversals

WASIM AKRAM PLAYS CRICKET like there's no tomorrow. And at times, it must feel like there won't be. He has already retired from cricket once, been hectored to do so on several other occasions, and suffered the sack as captain and cricketer perhaps as often as any player in history. He has been accused of cricket's most heinous crime, and escaped the supreme sanction only because of his accuser's tergiversations.

Then there are the injuries and infirmities. Shane Warne has likened his own life to a soap opera; Akram's is more like a medical drama. At one time or other, every pivotal point in Akram's body has buckled: groin, intercostal muscle, shoulder, pelvic bones. Then there have been the hernias, appendicitis, and diabetes leading to deteriorating eyesight. The money in Akram's family came from a business in spare parts; he could, over the years, have done with a few himself.

Few careers, then, have been clouded by so many intimations of mortality. But few will have such ongoing impact on the techniques of the game: Akram has been the most accomplished practitioner of a skill probably older than imagined, but formally acknowledged for little more than a decade. The development of reverse swing, as much as the renascence of wrist spin, was the headline trend of the 1990s.

Most cricket fans now have the gist of reverse swing, if not a grasp of its arcane physics: how ballast and wear on one side of a cricket ball achieve, in reverse, effects like those of protection and polishing. But they underestimate its subversiveness. Like the googly, B.J.T. Bosanquet's *jeu d'esprit* a hundred years ago, reverse swing was an act of counter-intuition, requiring dry not overcast conditions, extreme pace not 'time for the ball to swing', and the ball's deterioration rather than its preservation.

It has permanently altered the Test match ecosystem, emancipating the fast bowler with the old ball in the overs of an innings previously the preserve of slow- and medium-pace bowlers, and encouraging speed at a fuller length than was popular in the nasty, brutish and short 1980s. Particularly altered has been the predator-prey relationship between pace bowling and tailend batting. It seemed twenty years ago that, with the opportunities afforded by professionalism to rehearse secondary skills and the enhancement in protective gear including the helmet, tailend batting would probably improve in the long term; certainly, the real duffer became a comparative rarity. Helmets, however, afforded no protection from late-swinging deliveries speared at the crease line: an art with which Akram is synonymous.

Scyld Berry's theory is that reverse swing, by shortening the average duration of tailend innings, has been decisive in reducing the proportion of Test matches drawn since 1990. The view is persuasive although hard to test. What can be demonstrated is Akram's departure from earlier conventions of pace bowling. A greater share of his wickets – fifty-four per cent – have been bowled or lbw than any fast bowler of the last thirty years save his great rival Waqar Younis (fifty-seven per cent). For the purposes of comparison, Curtly Ambrose and Courtney Walsh obtained only a third of their wickets without the aid of a fieldsman or keeper, while the figures of Dennis Lillee (thirty-one per cent), Richard Hadlee (forty per cent) and Malcolm Marshall (forty per cent) imply the different devices of an earlier generation. That twenty-nine per cent of Akram's wickets have been secured lbw is freakish, considering the onus on a left-arm bowler seeking an umpire's indulgence from over the wicket.

Akram did not design his action with reverse swing in mind, but it proved close to ideal, with the fast arm and firm wrist imparting the necessary pace, and the seventeen-pace approach allowing him to sustain the effort involved in maintaining a consistently full length. No left-arm pace bowler since Sir Garfield Sobers has varied his angles as resourcefully, and not even Sobers was as skilful from round the wicket.

68

The technique itself is actually one of cricket's great wonders, defying all the usual injunctions of coaches to perfect a balanced run of gathering speed and a smooth action of seamless grace. After a breakneck sprint, he barrels through the crease, front foot pointing down the pitch, back foot towards the sightscreen, arm a blur. That he has been able to repeat this almost 40,000 times in international cricket beggars belief. Add to this the burden that Akram has borne as a batsman – he has almost 3000 Test runs to go with his 414 Test wickets – and one is compelled to consider another aspect of Akram's historical significance: his sheer durability.

The brunt of Akram's cricket has been borne by his groin and shoulder. His groin was first operated on in 1988, and again two years later. The latter operation was complicated when an adductor muscle separated from his pelvis, leaving his left leg only half as strong as his right: it was restored only by intensive physiotherapy. He first experienced shoulder pain seven years ago, while representing Lancashire, and delayed surgery, only to break down when he tried to bowl a bouncer during the Singer Cup Final in Sharjah in April 1997: there were further operations, a six-month lay-off and a regime of painkillers.

One could go on, but it all grows a bit gruesome. We might think instead of Akram as embodying the impact of medical science on cricket, both in terms of prolonging careers and facilitating modern schedules. Once upon a time, a single serious injury spelt more or less the end of a career. Players who recovered were hailed almost as miracles: think of Denis Compton's knee, Richie Benaud's shoulder and Dennis Lillee's back. Surgery today, by contrast, is almost as routine as the drinks break. There is far more discussion – some profound, some silly – about cricket being a game of 'mental strength' (something Akram also has covered, having been married for nine years to a qualified psychotherapist).

Akram's ultimate place in his country's cricket history is hard to guess. His star has waxed and waned. He was one of ten Pakistan captains in a bizarre period between March 1992 and August 1995, out of which also arose the allegations that resulted in his being levied a fine on the suspicion of involvement in match-fixing after the Qayyum Report. Political contacts have kept Akram going as surely as surgeons. Watching Akram in June, blasting out Adam Gilchrist and Ricky Ponting from the first three deliveries of the match in Melbourne, then looting an unbeaten 49 from 32 balls in the Gabba game, it was hard to escape the sad sensation that we in Australia might be seeing him for the last time. Then again, we've had that feeling before.

Pakistan Cricket Annual 2002–03

Steve Waugh
World Waugh

SOMETIMES, STEVE WAUGH FEELS, people get carried away with the importance of his office. Last November, for instance, newspapers sought the views of Australia's fortieth Test captain on whether his country should become a republic. Crikey. As it happens, Waugh voted in favour of an Australian head of state, though not from any heartfelt predisposition, and not expecting his view to be regarded as influential. 'It didn't worry me,' he says with a wry smile. 'I'm just an average bloke. Then all of a sudden I'm in the papers giving my views about it. Who cares? I wouldn't care.'

We're talking in the foyer of a Brisbane hotel a couple of days after Australia's victory in the First Test: its eleventh in succession. These are the good times, when the uncommon bond that his players have formed holds fast. All the same, he comments, it can still be a funny sort of a role: 'I used to look at Allan Border and Mark Taylor and think: "Good job, being captain. You might get a bit of stick when you lose, but there's a lot of credit when you win." But it's a real full-time job. You've got to have an opinion on everything.'

I'm also seeking Waugh's views today, though purely on the canvas of cricket. And it's surprising, really, that at a time when the game finds itself the focus of so much unwanted publicity and waist-deep in the views of others less qualified, these have not been solicited in greater depth. After all, in cricket terms, Steve Waugh is the guvnor: 129 Tests and 300 one-day international caps behind him, leader of the most successful and admired XI of his generation. But most of the time, his views are shrunken to soundbites and quick grabs. Perhaps observers have baulked at seeking his detailed views about the state of modern cricket because of that dead-pan demeanour, that personality so obviously self-contained. Perhaps they imagine his opinions to be conventional or predictable. If so, they couldn't be more wrong. Drawing on fifteen years as a Test cricketer, his thoughts are both trenchant and evolved, quietly but firmly held. He has a steady gaze, and seldom pauses, as though punctuation is for wimps. But this isn't haste – it's evidence of forethought.

A stray observation in the post-Test press conference a couple of days earlier seems a good place to start, Waugh having ventured then that international batting standards seem to have slipped. With the benefit of his historical perspective, what does he think about the fact that so many Test matches now fail to go the distance? 'The bowling round the place is pretty good,' he says. 'Most teams have got one or two good strike bowlers. But the batting, I think, isn't quite as good as it was. People aren't really prepared to bat long periods these days, or that's what it seems like. Indian players still get big hundreds but that's generally on their own soil where the pitches are flat and the outfields are quick. I think the batsmen are there, though, and it's good that teams are playing to win, which has been a big improvement in Test cricket over the last ten years.'

Even if that means three-day Tests? 'Three days and a result is better than five days of boring stuff.' What about a good old-fashioned draw? 'There's something to be said for a good drawn Test match, but I saw the recent one between England and Pakistan, and I was quite amazed at the commentary saying how well England were going and what a great Test match it was; it was the most boring Test match I've ever seen. They didn't give themselves a chance to win, which I can't understand. I'd much rather be playing the way we are.'

What about the paradox that countries nowadays tour each other so often – which should by rights make foreign conditions less intimidating – and the fact that home-ground advantage is still appreciable? 'You're always more comfortable in your own surroundings,' Waugh says. 'You know the conditions, you've got family and friends around you. Sometimes people have doubts about what's going to confront them as well, with different countries and cultures. That's something we've tried to change over the last three to five years. The way we tour now, we try to go out and see people, have a look at the place and take in some of its culture. Teams get into trouble overseas when all they do is sit around in their hotel rooms. You've got to enjoy touring or it's going to become a chore. I think that's happening to the West Indies side at the moment. You rarely see them outside the hotel. I think they should go out and check the place out a bit more, get involved and get their minds off cricket. I really think that's been the secret of my success away from Australia, that I've learned to go out and enjoy the places I've been, which means that you're not so focused on how homesick you are or what your form is like.'

Had I a daily deadline to meet, the interview could almost finish now. That's two headlines already: 'Lahore Test "most boring I've seen" – Waugh'; 'Waugh's advice to Windies: Give room service a miss'. Yet this would be misleading. Waugh is not a professional controversialist, merely a thoughtful cricketer who doesn't believe in doubletalk or fake sentimentality. We talk about how the brevity of modern tours may tax players acclimatising to unfamiliar conditions, and I ask if he's disappointed that the 2001 Ashes tour has been so truncated. His reputation might be as a traditionalist, but Waugh pronounces himself wholly in favour: 'I think the Ashes tour was too long anyway. The last couple of tours, the county games were a waste of time anyway. You were playing teams half full of Second XI players, you went through your paces and won quite easily, and that to me doesn't make a good tour. You want quality games. I hope that this time, with fewer games, counties will understand the importance of putting their best teams in the field.'

Waugh brings the same approach to his meditations on matchfixing. The old ways have not worked; a more centralised and genuinely global strategy is required, with strengthening of the International Cricket Council paramount. 'It's absolutely ridiculous the number of people they've got working for them, trying to run a worldwide game,' he says. 'They need ten times that amount of people. I think they're going about things the right way now. But the problem is that, with so many issues to cover, they can only scratch the surface of one before they have to move to the next one.'

That would entail, would it not, the national boards of control who so jealously guard their sovereign rights surrendering some of their powers to Lord's? 'That'd be nice,' answers Waugh. 'A lot of the time, it seems that the ICC's hands are tied; they can't enforce anything, not even penalties for match-fixing. It would be nice if they were able to state that *these* were the penalties, and if there was some ruling body that could make a firm decision so people know where the line in the sand is.'

Does that mean that the penalties the United Cricket Board of South Africa imposed on its malefactors were too lenient? That's a hard one, Waugh agrees; he retains a pang of fellow-feeling for Hansie Cronje, and would not like his exile to be permanent: 'Even people convicted of murder can expect to be out on parole in fifteen years. I'd like to see him involved in the game in some way, even if it's simply to warn younger players what might happen.' He can also find some sympathy for Cronje's co-conspirators Herschelle Gibbs and Henry Williams: 'I felt for a guy like Gibbs. He was pretty vulnerable. Cronje was his role model, and obviously pressured him into certain things.' The sympathy, though, isn't unconditional: 'By the same token, someone's got to be made an example of. You can't put everything down to lack of experience or lack of knowledge; so no, I didn't think the penalties were harsh enough.'

As for Australia's own record over the past few years, Waugh finds it hard to believe that the stain on his brother Mark and Shane Warne of having provided pitch and weather information to a bookmaker more than six years ago has proven so indelible: 'I don't see what they did as match-fixing. It didn't affect the outcome of a game. What they did was wrong, and they've said that they were stupid plenty of times; you can't keep being crucified for the same breach.' Waugh's disappointment with the press coverage of the affair extends further, too, to the freedom with which journalists have spread the contagion, idly diagnosing matches as 'suspicious' wherever events have not unfolded to type: 'You can't just pick a game out of the air and say: "There was something wrong there." That's why people watch sport. It's unpredictable. The underdog can win. Things happen in pressure situations. You have to expect the unexpected. I think it's just some journos cashing in and trying to make a name for themselves. I look back on my career and there would be twenty-five or thirty dismissals where someone could've said: "What's he doing there? Must've been taking money." I got a full toss on leg stump from Dan Marsh in the Pura Cup recently and hit it straight back to him. I've played 130 Tests. But that's what happens. Nobody's perfect.'

One suggestion for placing cricketers above suspicion – that of increasing match fees and prize money so that inducements from bookmakers will fade by comparison – meets similar scorn from Waugh. 'People in the past have offered us incentives for winning,' he says. 'And I've always said: "It doesn't matter if it's a million bucks or one cent. We'll still play the same way." That's the culture of Australian cricket, and I hope every country sees it the same way. If the players deserve that money because they've generated income, then sure, they should receive it, but I think it's a poor way to look at the match-fixing problem. If it's in your make-up or you're vulnerable to it, it's going to happen anyway; I don't think that increased prize money or wages would solve anything.'

At the same time, Waugh is sensitive to the income inequalities that exist across the spectrum of the Test-playing countries. At the recent meeting of international captains during the ICC Knockout in Nairobi, he even came up with the novel proposal that Test nations should agree on a base wage for players. Since meeting and talking with the Zimbabweans during Australia's inaugural Test in Harare last October, Waugh has been sorry to see players like Neil Johnson and Murray Goodwin renounce international cricket in favour of the first-class game in South Africa and Australia respectively. 'It's a vicious circle,' Waugh says. 'If you don't pay the players properly, you lose the players, the standard falls, you don't get the crowds and you don't get the income. Someone has to take responsibility for ensuring that standards in international cricket are maintained; whether that's the ICC or the boards. There's enough money in the game to ensure that countries like Zimbabwe are able to pay the players reasonable wages to keep the teams strong.'

This is of a piece with Waugh's concern that authorities should support the game at its newer frontiers, especially in Africa. 'It seems strange that you would get these countries involved and support them up to a certain level, then cut them off and let them go their own way,' he says. 'Kenya, for instance, have been admitted to one-day international cricket, but they haven't had a game for nine months. How are they supposed to improve and stay at that next level? You've got to get behind these countries and give them the resources they need. They've got some excellent players, Kenya; they could easily play Test cricket in five years if they get the right support, but they'll go nowhere if they don't.' Waugh, in fact, recently tried to interest Australian district clubs in sponsoring a group of Kenyans to play for a summer here; unfortunately, without success.

Waugh's words sit incongruously with the popular perception of Australia as the team that not only cares nought for its opposition but leaves their ears bleeding after verbal bombardment. But then, Waugh finds this image difficult to comprehend. To him, the Australian brand of cricket is unrelenting but entirely fair. Nor does he accept that the new laws needed their sanctions on verbal aggression. 'It's ridiculous,' he says. 'That's what the umpires are there for. In fifteen years, I don't think it's changed at all. I go back to grade cricket and the talk is three times as bad as what you get at international level.' Sledging, he believes, is an overestimated factor in Test cricket. As he puts it succinctly: 'It's not going to affect the good players, and the players who won't do that well you don't need to bother about.'

The constant and gratuitous reflections on Australia's on-field manner gall Waugh, peeling the gilt from the team's substantial achievements, which he would like recognised in full. His faith and pride in the XI he leads is unswerving, in private and public, the latter reflected in the string of tour diaries he has published over the last seven years. Waugh's eighth book, a journal covering Australia's all-conquering 1999–2000 season, appeared in November, entitled *Never Satisfied*. Just as he wouldn't want to be considered a fount of wisdom on the optimum Australian constitution, Waugh makes no great claims for his books, although he has no doubt that the effort involved in their composition has been personally enriching.

'The first five or six diaries,' he says, 'I was writing an hour and a half every night. Which is not an easy thing to do but I really enjoyed it. I knew if I could do that, the discipline would flow in to other aspects of my life. I'm proud of them actually. I haven't had anyone else do them for me. I mean, Geoff Armstrong [his publisher at HarperCollins] has been great – without him they wouldn't be possible – but I've put in a lot of work. And while it's really just a bit of a fun thing, you know, I think there's some good stuff in there; sometimes I'm a bit surprised that people don't pick up on some of it.'

Quite so. In fact, having recently re-read it, there's been something I've been meaning to ask Waugh about his first diary of the 1993 Ashes tour, in which he writes of betting £25 on himself at 8–1 with an English bookmaker to top the Australian aggregates during that series. He's quite pleased. 'There you go,' he says. 'There's stuff like that.' Then he smiles warily. 'You know, I can't remember doing that. But if you say so. Did I actually put money on myself? I'd better check that out.' For those interested, it's on page nineteen. They were innocent days, were they not?

Wisden Cricket Monthly January 2001

Steve Waugh
Undefeated

No Regrets. Never Satisfied. Never Say Die. Over the last few years, the titles and themes of Steve Waugh's enduringly popular tour diaries have explored a strangely similar set of ideas – of negation, prohibition, even downright refusal. Was there in the works, one wondered, a book called *Never Retire*?

Waugh would never have made so bold. Yet he has been a poignant sight in recent seasons. He has stirred from his team cricket of an increasingly hectic and relentless brand, as though at the head of a Zulu *impi*. But he has also been on his own path, each innings a little essay in resistance: of age, of rivals, of critics and of selectors, as well as of opponents. His individual performances have become a story within the story of Australian cricket, a subplot assuming proportions menacing to the triumphant principal narrative. The foreshadowing of his retirement on 26 November came, therefore, as a relief as well as a sadness, freeing admirers to contemplate what he has done rather than dwell on what he might.

In fighting on, of course, Steve Waugh was doing what he always had. The best of Waugh's innings had tended to involve holding on rather than leading a charge. Australia was 3/73 when he began his Test best 200 at Sabina Park in May 1995; 3/42 and 3/39 respectively when he began his two-part solo, 108 and 116, at Old Trafford in June 1997; 3/48 when he wrested control of the World Cup Super Six game against South Africa at Headingley in June 1999. 'In many ways, the delicate situation has always been the one I play my best cricket in,' he writes in his latest book – a remark akin to Menuhin saying he quite liked a bit of a fiddle. Waugh seldom, moreover, abandoned any innings prematurely. It used to be an old cricket truism that batsmen were vulnerable immediately after scoring centuries, relaxed and dis-armed by the landmark. Steve Waugh scoffed at the idea. His average Test hundred, inflated by undefeated innings, was 255 – greater than any other batsman, Bradman and Tendulkar included. A memoir of his batting alone would be called *Never Enough*.

This was not always the way of it. When twenty-year-old Steve Waugh made his Test debut at the MCG against India on Boxing Day 1985, he was chosen for his eye-catching strokes in the lower middle-order, handy change bowling and, above all, youth. It was an invest-ment in style, with the hope of substance, that did not fructify at once. After fifty-two Tests – a career the length of Bradman's – Steve Waugh was averaging in the mid-30s with the bat and more than 45 with the ball; without three fruitful Ashes Tests in 1989, moreover, his batting average shrank to less than 30. He had already spent one sea-son, 1991–92, by the wayside; he had not, the following summer, made the number three position in the Australian XI quite his own. It's arguable that Dean Jones, who in his own fifty-two-Test career was averaging almost 47, would have been the better bet in England ten years ago; disgruntled Victorians muttered darkly about the New South Wales hegemony in Australian cricket.

It was this 1993 tour, however, that marked the turning point in Waugh's career. That it was the first on which he kept a diary – an idea with which he came up himself rather than having it put to him by a publisher, and to which he applied himself with scrupulous care – strikes me as more than a coincidence. A calculating cricketer had always lurked beneath Waugh's natural talent; from childhood, for example, he had always counted his runs while batting, and never lost the habit. Now that cool, rational, meticulous cricketer began to crystallise. A glimpse of the Waugh to come was seen at Headingley when he batted most of the second day in partnership with his cap-tain Allan Border. 'Batting with Border always makes you concentrate that little bit extra,' Waugh wrote, 'because you can see how much it means to him to give his wicket away.' He admitted to having con-sciously 'set my sights on getting a big hundred' – there were, he had come to understand, 'so many tough times in cricket'.

Border was 175 and Waugh 144 at the close, but the former surprised the latter by batting almost another hour the next morning, explaining his desire to 'cause further mental and physical disintegration'. 'Mental disintegration', of course, is the expression Waugh has used in recent years to describe the psychological pressure his team exerts; Border is, perhaps, its intellectual godfather.

In some senses, too, it was the end of Border's career in March 1994 that prepared for Waugh the role he would fill: that of the immaculate bulwark. Under Mark Taylor's captaincy, Waugh turned five Test centuries into seventeen. When his twin brother Mark was at number four, Steve's sternness at numbers five and six seemed particularly pronounced: Mark batted as if in a dream, Steve as if in a trance. After watching his 170 and 61 not out at Adelaide in January 1996, indeed, two Sri Lankan players asked Waugh if he meditated; he seemed, they insisted, to be on a different plane of consciousness while batting.

This entailed sacrifices. Some of his boyish brio disappeared; he ceased to hook and seldom pulled. Again, though, this brought out qualities in him rather than suppressing existing gifts. He had, in that pinched face, those gimlet eyes, that low-slung stance, the elements of an aura – and he used it. He would come out to bat quickly to 'own that space in the centre'; he tugged his old Australian cap low, as if to draw strength from the tradition of which he was part. His back-foot drive was executed like a sock to an opponent's jaw, his slog-sweep like a haymaker to the solar plexus.

That self-absorption was held against Waugh when Taylor announced his retirement in January 1999. Neither his succession nor his success were preordained; when Australia drew Test and one-day series in the West Indies that they had been expected to win easily then went to the brink of exiting the World Cup, his aptitude seemed in doubt, not to mention his attitude. When he had Australia inch to victory over the West Indies at Manchester in an unsuccessful effort to exclude New Zealand from the tournament's Super 6 stage, he was coldly dismissive of criticism: 'We're not here to win friends, mate.'

Australia, of course, won the Cup – and Waugh, architect of that triumph against South Africa at Headingley and Edgbaston, so much more. Waugh's attitude to winning friends underwent a subtle transformation too. The first group were in his own dressing room. Waugh as captain reminded me of a resolutely self-contained man surprised to find a new world opened to him by fatherhood – which, perhaps, he was as well. From the rises of Brett Lee and Adam Gilchrist, and the resurrections of Justin Langer and Matt Hayden, he derived as much satisfaction as from any personal landmark; from instilling in his team his own ethos of continuous improvement, he created a culture that outlives him. Remember when you were a kid, looking up to and wanting to join the toughest gang in town? This Australian team is that gang. It has its own codes, creeds and customs. It assimilates newcomers, largely because they wish to be assimilated. And it speaks with one voice, that of its captain. As Stuart MacGill once put it, its prime directive has been: 'When in Rome, do as Steve Waugh.'

Waugh has also acted like the peasant who accompanied Caesar's men on their triumphal processions through the streets of Rome whispering: 'Remember, you are dust.' During the Test in Hamilton in March 2000, for instance, Waugh donned the cap he had worn during his first series against New Zealand, when Australia had been beaten – a warning against complacency. Likewise before the Edgbaston Test in July 2001, he gave a brief but heartfelt address about what it was like to lose an Ashes series – something only he could recall. Waugh's captaincy was almost – but not quite – a personality cult. Not quite because the loyalty of Waugh's players reflected his loyalty to them. The Australians' team psychologist Sandy Gordon once asked team members to complete a questionnaire which included identifying scenarios they found 'particularly mentally demanding'. Waugh replied without hesitation: 'Selection meetings – having to leave out players.'

In one aspect of the Australians' approach, too, Waugh was able to win goodwill as well as games: through 2001 (3.77), 2002 (3.99) and 2003 (4.08), his team's scoring rate increased, until it seemed to be playing cricket on forty-five rpm while opponents remained grooved at thirty-three rpm. As *Inside Edge*'s Chris Ryan has commented astutely, the effect has been a kind of cricket with only two possible outcomes: victory or defeat. On only five occasions in Waugh's fifty-three Tests have his teams been stalemated. Guaranteed an outcome, Australian Test audiences have risen from 411,335 in 1999–2000 to 568,324 last summer; everywhere they go now, Australians are feted for their enterprise.

In another respect, sadly, Waugh did not quite deliver on his promise. For one who prided himself on his historical awareness, he condoned in his team a very modern truculence. This did not do him justice: he was an exceedingly generous opponent off the field, and an uncommonly charitable man away from the game, as evidenced in his patronage of Nivedita House orphanage in India. Yet here, Waugh's innate resistance, his ability to shut out doubt and discord, became a weakness: he failed, in this case, to harness his stature for the good of all cricketers.

For his stature grew and grew. In the final stages of his career, Waugh reminded me of Swede Levov, the Jewish athlete in Philip Roth's *American Pastoral*. 'I have to tell you that I don't believe in death, I don't experience time as limited,' Levov says. 'I know it is, but I don't feel it.' His final Test appearance in England after a severe calf injury seemed to symbolise his indestructibility: an unbeaten 157 on one leg, like an old man toying with children in his backyard. The summer of 2001–02, however, was a *memento mori*. In nine Tests against New Zealand then South Africa at home and away, Waugh cobbled together only 314 runs at 24. A few years earlier, such a streak would barely have occasioned comment; in a thirty-six-year-old, it was interpreted as the last beats of a fading heart.

Waugh's Test captaincy was protected by the selectors' Napoleonic faith in lucky generals, but the form lapse cost him the one-day job. This, Waugh told Greg Baum of *The Age* recently, was a greater blow to him than he realised at the time: 'You put up a brave front and say you're going to get back in there and fight hard … At the end of it, I was at the crossroads, wondering if I should retire or play on.' A timely century against Pakistan in Sharjah in October 2002 silenced doubters; his stunning hundred against England at Sydney a year ago turned them back into stark raving fans.

It passed without remark at the end of November that Waugh's designation of 2003–04 as his valedictory summer is without Australian precedent. Ian Chappell, Border, Taylor, Healy, Slater, Geoff Marsh, Waugh's own brother Mark: none of these sought or were rewarded with last farewells. Greg Chappell, Lillee, Marsh, Boon: these had a Test to say goodbye, but either wished or were granted no more. And though Bradman received his three cheers, nobody knew if the innings would be his last. But then, Steve Waugh has been a cricketer *sui generis*. The future biographer even finds their title ready-made: *Never Again*.

The Wisden Cricketer January 2004

Shane Warne
Positive Spin

'IT IS SOMETIMES NOT A BAD THING for a professional sportsman to sit in the crowd and watch from the other side. It is a reminder of how much you miss it when it's not there.'

When Shane Warne confided this reflection in *My Autobiography* in 2001, it was with regard to past incidences of involuntary spectating rather than an expectation of a future exile. But at the end of perhaps the most frustrating year of his career, during which he has seldom been fitter and never less occupied, it is too tempting not to ask him how the thesis stands up.

Pretty well, he reckons: 'I did write that. And it is true. When you're watching, in fact, you tend to pick up other things. Sledging. Style of play. Appealing. Body language. Which, when you're involved in the game, you're less aware of, because you're focused on the match and your opposition; because you're emotional. Some of the things I've seen I've learned from. I might even have been guilty of them a few times in my career.'

Not, Warne admits, despite a season behind the microphone in the Channel Nine commentary box while suspended for the heedless popping of a banned diuretic, that he is one of nature's watchers. 'I've enjoyed the commentating,' he says. 'They've all looked after me in the box, and I think I'm quite good at it. I mean, I do love cricket, and I've watched more of it than I ever have this summer. But I'm not someone who has to watch it all the time. If someone's been bowling a good spell, one of the quicks, or MacGill, or Katich, or one of the batsmen have gotten on top, I've watched it pretty closely. But I prefer doing it. There've been times when Zimbabwe and Bangladesh have played this summer when I've had a bit of a snooze.'

Watching the World Cup, which should have been the crowning glory of Warne's limited-overs career, was particularly difficult; sitting up late at night to follow his mates in the West Indies a little less so, though still frustrating. But Warne operates in a negativity no-fly zone. 'I'm looking at the positive side,' is a phrase he scatters freely in conversation, and he manages to make it sound like a creed rather than merely a cliché.

The loss of a year? Think again. Warne regards it as the gaining of two – at least – on the end of what might by now have been a completed career. Positive person? Warne agrees he is: 'Maybe that's why I play cricket the way I do.' Perhaps it's intrinsic to leg-spin. Remember how Richie Benaud was nicknamed Diamonds, because 'Dusty' Rhodes thought that if he put his head in a bucket of shit, he'd come up with a mouthful of gems? Warne would find at least a pile of casino chips.

Even the indiscretion that cost him the year has been rationalised away. Warne dismisses it as 'not checking a book' – the book that would have told him that the fluid tablets he took had the 'potential' to act as masking agents for steroids – with the implication that checking a book is something it's a bit hard to expect of people.

The only question that gives him pause, even momentarily, is what might have happened had he received the two-year ban many thought condign: 'Not sure. Don't know. What I would say is that, if I'd been guilty of trying to hide something, if I'd taken the diuretic to cover up a performance-enhancing drug, I should have been rubbed out for life. But I didn't.'

So is there anything he has missed? The answer is interesting. 'I've had a fair bit of contact with the guys over the last year. Darren Lehmann, Punter, Damien Martyn, Binger, Haydos. In fact, I'm catching up with Punter tonight. But it's different to when you're playing. You know the thing I really enjoy about walking onto a field for a game – any game, whether it's a Test, or a one-dayer, or a club game? It's that you don't know what's going to happen. You don't know whether you're going to knock 'em over, or you're going to get slogged, whether you'll have a bad day, or take a couple of screamers.'

The baggy green? 'Yeah, don't get me wrong. That means a lot to me. But what I miss when I'm not playing is that unpredictability.'

What he would not miss is the unpredictability of his personal life, which reached something of a zenith last August when he was accused of harassment by a South African model and of infidelity by a Melbourne stripper. 'He told me he had an open marriage,' said the latter on television. 'I asked him about his wife and he led me to believe they weren't together.' And for a time, they almost weren't. Simone, whom Warne married in September 1995, took their daughters Brooke and Summer and went to live with her parents; Warne was left with his son Jackson in the palatial 85-square residence that he has spent almost three years turning into a cricketer's Xanadu, complete with cinema and popcorn machine.

Warne is less positive and forthcoming where his marriage is concerned. 'Marriage is a tough job at the best of times,' he says. 'Simone and I have been through a lot together, most of which, unfortunately, has been my fault. But that's in the past now, and we're looking forward to getting on with our lives.' If he doesn't invite you further, that's because the media has tended to welcome itself into Warne's personal life regardless; the extent of intrusion last year startled even a hardened campaigner like Warne. On one occasion, Warne and his wife agreed to meet for a conciliatory talk in his car in Port Melbourne; within ten minutes, two Channel Seven vehicles containing film crews had pulled up on either side, boxing them in. 'I still don't know how it happened,' says Warne. 'If you hear anything, let me know.'

Publicly, mind you, Warne handled this imbroglio perhaps better than any other in his career, maintaining his silence and his dignity while all about him were losing theirs, and at the height of hysteria rather shrewdly disappearing to Spain and the UK. Whose idea was that? 'Mine. Well, both of ours. I said: "Where would you like to go Simone?" She said: "Spain." I said: "Let's go."'

It proved a tonic. It turned out that, in this affair, being suspended had advantages, there being no need to create a pretence of normality and continue playing cricket. Observers were also relieved of the unedifying spectacle of Cricket Australia acting as a moral arbiter. Where Warne's indiscretions with a nurse three and a half years ago cost him the Australian vice-captaincy, Cricket Australia CEO James Sutherland this time offered support: 'There are obligations and standards incumbent on players, but there is also a boundary between their public lives and their lives as private citizens.'

Warne appreciated Sutherland's sensitivity; indeed, he thinks highly of the administrator, who is, of course, a former Victorian team-mate. But the irony of the situation was not lost on Warne, who commented tartly in *My Autobiography* that Cricket Australia's directors paid too much attention to the media, and that the relationship between sins and first stones might profitably be explored: 'I wonder what might happen if the backgrounds of the fourteen directors themselves were investigated. If the same rules apply, as they should, then they, too, must have led unblemished lives or they are not fit to do their jobs.'

'Well, maybe they read my book and took the advice,' Warne jokes now. 'But yes, one minute my personal life was their affair; next minute it was none of their business. Then again, there wasn't really much they could do. I wasn't vice-captain anymore. I was a contracted player but the contract was suspended and I was receiving no money. If that had been their attitude in 2000, then I might have stayed vice-captain and things might be different today. Who knows?'

Who does? If the rigour of his regimen is anything to go by, Warne is certainly planning to be around for a while. In line with Terry Jenner's advice, he did not recommence bowling too early. But since he resumed training on 27 December, his routine has involved three bowling sessions a week of fifty deliveries each, bike rides, swimming sessions, sprint work, weight programs, skipping, boxing and karate. His cross-training has even involved kicks of the football with his friend Aaron Hamill of the St Kilda Australian Rules club, where Warne played in the reserve grades fifteen years ago and which over the years has been a favourite haunt.

Nobody seriously doubts that Warne's old Test berth is his for the taking, and that he will form part of the team Ricky Ponting leads to

Sri Lanka for the tour beginning on 8 March. It might be a year since his last game, but it is only two years since perhaps his best: that hundredth Test in Cape Town where he extracted eight wickets from 98 overs, scored 63 and 15 not out, and won it almost single-handedly.

The question is not so much about his place in the Australian team as it is about his role in Australian cricket. Times are changing. Warne may have extended his career, but under what circumstances will it be played? On more than one occasion, Warne has said he would have enjoyed playing for Australia during the 1970s. He has been thinking primarily of the scrutiny of players' lives, but there is also the issue of the way the game itself is heading.

Warne's approach to cricket suggests a player not completely at ease with Buchananite precepts. In *My Autobiography*, he confessed a preference for using 'the old brain' when he played; computers could only be a 'backup'. He now goes a little further: too much analysis of cricket can be harmful. 'When you come off the field, I think, you have a bit of a chat, think about the things you got right, what you might do differently, then you move on, and you pick it up the next day. The danger is that if there's too much analysis, people get confused, or bored, or switch off.'

His admiration for the encyclopedic but intuitive Ian Chappell, with whom he is now tight enough to have played a seven-hour game of backgammon in Darwin during the Test against Bangladesh, is unstinting. 'We check up on him,' says Warne. 'We'll look up a game he's talking about, and once or twice what he remembers as a 50 turns out to be a 15, but we hardly ever catch him out. I've never known anyone with his memory for games.'

Warne reckons his mindset is similar. 'I'm pretty good. The one-dayers, they get a bit blurred. I've played seven against South Africa in Johannesburg, for instance, and I have a bit of trouble telling them apart. But I do think I read the game well. I understand it. I've enjoyed my opportunities as a leader. I enjoy the role of a senior player, and hopefully my experience can make a contribution to the team in future.' The opportunity to skipper Hampshire, too, was fundamental to his decision to take up the cudgels there again this season.

What about the end? When we digress, as two Melburnians are wont to do, into the *lingua franca* of Aussie Rules footy, I remind him of one of St Kilda's best games last season, the finale to the career of its indestructible Nathan Burke. The crowd's salute to Burke at the end of the game was definitely one of the more moving tributes to a player I can remember – the St Kilda fan next to me burst into tears, the Richmond fans who'd seen their team utterly thrashed solemnly shook hands with their counterparts, as though at a funeral.

Is this what Warne would like? Something simply spontaneous and demotic in front of a home crowd. Or would he prefer a Waugh-like pageant, in which each city has the opportunity to say farewell and vice versa? Excitement about his career's resumption, it turns out, hasn't curtailed consideration of its conclusion. 'What I would like is to finish on my own terms,' Warne says. 'I think I deserve that.' He doesn't believe that his lost year will tempt him to play more cricket than he should. 'There's plenty of it, though, isn't there?' he says. 'Too much, really.' But, at the moment, better too much than not enough.

The Wisden Cricketer March 2004

Cricket Tragics

R.W. Wardill
The Hero and the Ham

IN HIS ROLE AS THE PRINCE OF DENMARK, Walter Montgomery was a truly transcendent theatrical experience. 'Go and see Montgomery's Hamlet,' urged one critic. 'And then thank God that you have lived long enough to enjoy the opportunity.'

The Melbourne of July 1867 had already been somewhat spoiled for Hamlets, with the impresario George Coppin having imported tragedian James Anderson to pack his Haymarket Theatre, while Tom Sullivan had just completed 1200 tragic nights at the Theatre Royal. But Montgomery was fresh from London's West End, where his gibbering, foam-flecked impersonation had caused uproar. The storm centre shifted the moment the curtain rose on him at the Royal, the newspaper correspondence it provoked eventually filling a twice-reprinted book *Was Hamlet Mad?* Like Anderson, Montgomery then diversified, rotating the tragedies with such vigour that he and his rival occasionally performed identical plays the same night at their different venues.

This is not, however, the story of Montgomery in his great theatrical role. It is the story, strange but true, of his role as a benefactor of Australian cricket. A New Yorker who affected British mores, Montgomery probably felt constrained to like the game. Whatever the case, 125 years ago on Boxing Day, the man who put the ham back in Hamlet provided the stake for the first century in our first-class cricket history, at the Melbourne Cricket Ground. And, in doing so bestowed on a truly remarkable sportsman – Richard Wilson Wardill – the single memorial of an otherwise benighted life.

While there can be hardly a barstool in its rich Great Southern Stand not named for an athlete or administrator, there's barely a trace at the MCG of the man who first gave the ground reason to speak of a 'Boxing Day tradition'. Wardill should, by rights, be in lights. In an era when it was all waiting to be done, he went and did it. For eleven years a committeeman at the Melbourne Cricket Club, he 'civilised' a pastime that had been no more refined than mud-wrestling with bats. He thrashed out rules for Australian football as a member of the code's founding body, and was the first captain of what began as the Melbourne Cricket Club Football Team. And, devoted to the 'muscular Christianity' of amateur athletics, he helped found its first administrative body and organised the first visits from touring champion troupes.

Cricket prospered from Wardill's work even after his death. He proposed and propelled negotiations with Dr W.G. Grace which drew England's mightiest cricketer to the colony for the first and only time. His brother Ben was an astute administrator, too: he was MCC secretary from 1878 to 1911 and responsible for many of its greatest earthworks, not to mention its earliest innovations (like floodlit football in 1879). If, as they say, you seek their monument on Boxing Day at the MCG, look about you – just don't bother looking too long. For, while those who gamble and lose today are called entrepreneurs, those who gambled and lost a century ago were embezzlers. And if, in their shame, they tossed themselves into the Yarra, they were called nothing at all.

*

When Anthony Trollope assessed colonial mettle in 1871, he could have had Dick Wardill in mind. 'The Victorian old man hardly as yet exists,' he concluded. 'The men who have hitherto prospered best in Australia are they who came young from the old country, without much money, with great energy, and with a strong conviction that fortune was to be made by industry, sobriety and patience.'

Sobriety might have been stretching it, but the lad who left Liverpool in 1858 was certainly an apostle of animation in a teenage city welcoming newcomers like him at 250 dreams a day. Melbourne was a citadel of gold, and about its only infrastructure was pubs, of which Wardill would have found 135 in the central precinct alone. They amounted to an essential service: the average Victorian of the time drank seventy-three litres of ale a year, brewed by any of 100 breweries.

Within two years of arriving, Wardill was an honorary secretary of the Melbourne Cricket Club so diligent that he slept at its Richmond police paddock (aka MCG) to guard against vandals. The club had just come of age, which was something Wardill hadn't, although he kept that to himself. He was just nineteen, although he claimed throughout his life to be five years older, hoodwinking even the *Australian Dictionary of Biography.* By 1861 Wardill was captaining the MCC at the two-year-old game of Australian football, conceived for cricketers' calisthenics by the star all-rounder Tom Wills and still little more than a forty-a-side catharsis on a half-mile field.

Wardill's first important cricket engagement at the MCG on New Year's Day 1862 was also in grand manner against the first English touring cricket team led by Surrey's Heathfield Stephenson. There were 15,000 spectators, a band, a balloon and pre-match punting in nuggets – not to mention 500 cases of beer. Wardill was baptised a first-class cricketer by Victoria against New South Wales the following week, and two months later represented no less an outfit than 'The World': a grab-bag of Victorian and English players who took on a team made up of the Surreyites among the visitors at the MCG in March 1862.

National allegiances were then simpler to ignore than inter-colonial ones, for cricket often became a suitable civil war surrogate, and the following season brought matters to a head. A crowd of 6000 at Sydney's Domain rebelled when, after he had strayed from his crease at the calling of 'over', local batsman Syd Jones was run out by Victorian wicket-keeper George Marshall ... and Victorian umpire John Smith. Bad language stopped play after twenty minutes, almost provoking a lynching and advancing retirement plans for both umpires. Wills himself had to poll his team before electing to continue the match, although he had to 'surrender' when the dismal Victorian second innings reached 8/45 and two dissidents refused to bat. The colonies, in due course, refused to meet on the cricket field for two years, vocally supported by their newspapers. 'Scratch a Russian and you find a Tartar,' decided Melbourne's *Evening News*. 'Scratch a New South Welshman and you discover the evidence of convictism.'

As they kicked their heels, Wills and Wardill kicked more rugby balls and became deeply involved with Wills' cousin Colden Harrison in bringing order to the chaos of Australian football. Wardill in particular became treasurer of the MCC's Amateur Athletic Sports Committee that in June 1865 staked a Challenge Cup for Melbourne's match with Geelong, before on the Queen's Birthday leading the former side onto the football cordon roped off next to the MCG. Some sort of cup challenge seems to have occurred the night before. *The Australasian*, in the euphemistic language of the era, reported Wills as unexpectedly 'absent from illness', Wardill 'suffering from an indisposition' and another colleague as 'not kicking with his accustomed vigour' having 'in turf parlance, passed a heavy night'. Perhaps one needed a degree of prior disorientation to play. One boundary of the field was marked by the galvanised fence of the MCG, the other by gum trees, and casualties were numerous. That is when they weren't scraping themselves off another sixty metres of goal-side gravel.

After an even first half in the cup game, three-quarters of an hour 'refreshing the inner man' spurred a half-hour Melbourne goal glut: two. Melburnians 'kissed mother earth' as the Geelong men came back, but home advantage told. Out of sheer *joie de vivre* a scratch match was then played to entertain latecomers. A couple of months later the MCC retained its trophy by beating South Yarra in a match spread over two Saturdays, with Wardill again prominent in what *The Australasian* called the team's 'fighting brigade'. And he would be MCC's man at the May 1866 meeting in Swanston Street's Freemason's Hotel where 'Australian Rules Football' was belatedly given some Australian rules.

Wills and Wardill, meantime, had helped guide a rapprochement between Victoria and New South Wales. They had called a meeting at the Freemason's in February 1864 proposing a Victorian Cricketers'

Association to get the states sharing a field again and, after nineteen months of what *The Australasian* called 'a little coquetting and courteous correspondence', they did. Ironically, Wardill wasn't on it (indisposed), nor was he a starter for the sequel a year later (unavailable), although brother Ben was. He'd followed his elder to the colony four years later, also joining a sugar company, but his 4 runs in two innings in this solitary proper state appearance at the Domain in December 1866 probably flattered him.

Dick Wardill was free, though, for a notable MCG encounter on Boxing Day: he led MCC versus T.W. Wills and Ten Aboriginals. Wills had been hired by a Hamilton station owner to 'coach the darkies and bring out their best possible form' and, when this entrepreneur then sent MCC photos of his team 'in proper costume and quite respectable in appearance', urged the offer of a game. Among the 10,000 curious, *The Australasian* was bewitched by Aboriginal assimilation of cricket's 'nice points'. 'It has been proved that the blackfellow has an extraordinary readiness for picking up knowledge of the game, however deficient he may be in other respects,' it said. 'And such being the case, it is only reasonable to suppose that if properly managed and instructed the natives might have been turned to better account than has been the case ... they might have become civilised and respectable members of society ...' Wardill made the match's highest score of 45 before being bowled by Wills, who was becoming a convert to his team's cause – an amazingly wholehearted one too, given that his own father had been one of eighteen victims of a grisly Aboriginal slaughter just six years before in Queensland, and that he still packed a pair of six-shooters in case of trouble. But he was to be a disillusioned one when a planned Aboriginal tour of England in 1867 fell through – undone by an unscrupulous promoter and the bureaucrats of the Central Board for the Protection of Aborigines – and cost him a modest bundle.

*

Suddenly, times were tight. A VCA cash crisis loomed. In April 1867, the cricket correspondent of *The Argus* admitted what everyone already knew, that the association was 'a failure' and that 'its excheq-uer is at present empty'. Refinancing options had to be considered. With Montgomery and Anderson in town the theatre seemed licensed to print money, and Dick Wardill had a brainwave. Why not a charity evening at the Theatre Royal in October? Its nature was unabashed: the VCA could not go past Edward Bulwer-Lytton's comedy, *Money*, as a fundraiser. And Wardill probably singled himself out as its lead: the roguish Captain Dudley Smooth.

He certainly found few difficulties in the portrayal. *The Argus*'s critic thought him the 'best I have seen' in the 'Deadly Smooth' role, and 'quite as much at home on stage as between the wickets' even though the writer was far from scattering rosebuds (one thespian he singled out as 'conveying the impression a scion of the nobility had descended from a greengrocer', another as behaving like 'an absolute ninny'). All the more important to Wardill, however, was welfare the stage was about to offer. Just before Christmas and the next inter-colonial game he opened a letter from Edward Scott's luxury hotel (the drop-in centre for beautiful people across Collins Street from the Customs House) signed by none other than Walter Montgomery.

The letter survives in the records of the Melbourne Cricket Club even if the reasons it was written do not: with the Duke of Edinburgh, Prince Alfred, about to pay Melbourne its first royal visit, Mont-gomery may have been endeavouring to scatter a little largesse. He would, he explained, be delighted to entertain the colonial teams at one of his 'readings' (his 'name alone' would be 'sufficient passport to reserve seats') and to spice their subsequent match. 'I shall do myself the great pleasure,' he said, 'of presenting a little cup for the matter of the highest score during the match if you will permit me'. Wardill would permit him all right, and probably made mental note at the time that this was a game for which he'd better not be indisposed.

Wardill had never been a punctilious cricketer. MCC minute books show that, on odd occasions, they turned down his offers to play because he'd failed to turn up at a previous fixture. The previous season he seems to have forgotten he was captaining Victoria against sixteen Tasmanians at the MCG, and the players were left idling for about thirty-five minutes on the second day with no-one game to start in his absence (for the record he top-scored). But on Boxing Day 1867 he took first ball and, in contrast with his customary combativeness, played out four maiden overs before a first sober single. A trip to the crease in 1867 was a bit like a trip to the casino today. The odds were stacked with bowlers – round, under and overarm – on pitches of dust (before rain) and mud (after).

Wardill had entered a game in 1861 in which batsmen over the prior decade of recognised first-class cricket had scrambled just 2700 runs at 6.5 runs every time they went in. Lots of extras helped (wicket-keepers would be discretionary for a few years yet), but only two fifties had been saluted in the seventeen years preceding Wardill's Boxing Day brush with the New South Welshmen, both by English visitors.

Wardill spent four hours and twenty minutes in occupation this 26 December: a length of time in which club games had been known to start, finish, organise a bat raffle and turn into an impromptu game of football. Said *The Australasian*: 'Whatever good bowling there was he had his share, and the best share of it, and he played it well, never once being tempted to indulge his favourite propensity for having a slog.' Wardill scored not merely the first hundred in our history, but the first fifty as well; partner George Robertson scored the second, an hour later, and theirs was the first 100-run partnership in Australian first-class cricket history. The only feature missing was a visit from the visiting Prince – whom, it irked some, had headed for the vulgar turf that day instead – but the game in this country crossed a Rubicon by the Yarra that day 125 years ago. *The Age*'s breast swelled. 'It was said that our cricketers were effete, and that when we came to be compared to Sydney ... we would have again to hide our diminished heads,' the newspaper said. 'These must have lost sight of the fact that, however unlucky Victoria may have been on the turf or at the butts, she has always maintained her supremacy at cricket.'

The Australasian paused for some true connoisseurship. This was a match that repaid only morally righteous watching. Poltroons should find other distractions. 'Cricket,' it opined, 'is a game which ...

to thoroughly enjoy requires a person's undivided attention; and during such an innings as that of Robertson, Wardill or [NSW captain Nat] Thompson one's most intimate friend is often *de trop*, especially if when a brilliant cut is made he is retailing for your special benefit the latest canard about The Barb, or asking your opinion as to a "good thing" for the steeplechase.' It exulted too that betting seemed to have been banished, pre-match skulduggery was no more and that even the players appeared to be showing greater rectitude. 'There is luckily no anxiety at present about Tommy Wills "passing a bad night", nor is it necessary to shepherd any of our men previous to a grand match for fear of their being "nobbled".'

Wardill's head was particularly clear over the game's quite unprecedented four days (prolonged by a dust-storm, followed immediately by rain). His unbeaten 45 on the Monday saw Victoria home by seven wickets, a victory that seems to have quite dumbstruck MCC president Charles MacArthur as he handed Wardill a bonus bat and Montgomery's cup: he could not find a word to say. Fortunately, *The Australasian* noted, 'any deficiency on the part of the president [was] made up by the hearty applause' of the 600 at the post-match ceremonies. Never mind the Prince: the crowd had appointed one.

*

How that prince became knave is a tangled matter, but piecing the allusive and erratic news coverage of the next five years hints that life was never quite the same for Dick Wardill. Wardill may have concluded that a putative prince should develop the interests of a proper one, like the races. And card games. There was no lack of avenues for an adventure capitalist like him: trotting meetings had begun in 1860, the first Melbourne Cup was run the following year, and what was not a pub or brothel in Little Bourke Street was a Chinese gambling den. Wardill was lucky enough to stay interested – in one famous sitting raking £450 – but rarely enough to sustain his habit on an annual salary of £400. He had worked for Victoria Sugar almost a decade and may have figured it was time the company work for him.

The money from *Money*, as it were, steadily ran out. Wardill was forced to fold the Victorian Cricketers' Association, and surrender responsibility for the intercolonial game to the Melbourne Cricket Club in October 1868. But he remained, at least outwardly, the sort of respectable figure on which cricket could rely, taking on the responsibility for Victorian selection with Wills and D.S. Campbell four months later. He soon had to shoulder more duties. The behaviour of Wills – who had confessed to English relatives that he only endured the death of his father by behaving 'as if he were himself present and could approve of what I do' – had grown increasingly erratic, thanks to alcohol and financial embarrassments. When he misbehaved once too often on the eve of a game against Tasmania in February 1869 and was sacked as captain by his own team, Wardill filled the breach.

Wardill's form had been poor that season – indeed he would perish to the first ball of the subsequent game at the Domain – but his elevation was warmly greeted. 'We need not tell Mr Wills that of late cricket has fallen somewhat from its high estate … and for this change some cricketers are themselves much to blame,' lectured *The Australasian*. 'We do not wish to wound anyone's feelings in particular, but no doubt the XI had very good reasons for selecting Mr Wardill as captain and … as a gentleman of unimpeachable honour and integrity Mr Wardill has the entire confidence of the cricket community.' Wills remained the proverbial prodigal talent by winning the game with the ball, but the pair would play just once more together in the following year's MCG intercolonial. Even then they would have only supporting roles to Lieutenant Charles Steward Gordon – a British officer with the 14th Buckinghamshire Regiment of Foot briefly stationed in Melbourne after serving in the New Zealand wars – who scored a remarkable one-off 121 in his only game on Australian soil. Wills then broke down completely, requiring confinement in the Kew Asylum, and Wardill played increasingly with the air of a man preoccupied.

Not all Wardill's thoughts were vexing. In May 1871, he married Eliza Helena Lovett Cameron – sister of a team-mate – at South Yarra's St Stephen's Church; they moved into a small but presentable home in nearby Mona Place. But he would have been disturbed by another item of matrimonial news from abroad concerning his old patron Walter Montgomery. Hamlet had been literally spurned by his Ophelia (a local belle Marion Dunn) in 1868 when she, very cruelly for an actor, married a critic: no less a figure than Marcus Clarke, a twenty-one-year-old prodigy columnising in *The Argus* as 'The Peripatetic Philosopher'. Montgomery had quit the colony, headed for the US and Wardill would have read three months after his nuptials of Montgomery's own to an American actress in London. He would two days later have learned of Montgomery's truly Shakespearean demise: the actor retreated to his hotel, called for his favourite horse, and shot himself.

*

Misfortune was now trailing Wardill round his gambling haunts. By now he'd embezzled nearly £5000 from Victoria Sugar, and was having to play for larger and larger stakes to recoup his losses. A pall settled over his cricket. He took 27 wickets for MCC in the coming season, but his batting average shrank to less than 7 and he became embroiled in an infamous club game with East Melbourne known as the 'lost ball' match when the rivals' scorebooks failed to balance. He even lost here.

The only way out was a successful bet that would redeem all previous failures. And, at the Cricketer's Dinner he organised for MCC in March 1872, Wardill spoke excitedly of recruiting a touring side to be led by Dr Grace. The Doctor was just twenty-four, but his cricketing renown girdled the globe, and so many shared Wardill's enthusiasm that some £3750 was subscribed when he proposed at a meeting at Scott's Hotel in May that the Englishman be approached. Dr Grace, however, played for stakes that shocked even Wardill. He wanted £1500 plus expenses for himself alone, never mind the rest of the team, and at the end of the English summer settled for the less strenuous diversion of a cricket tour to North America. The disappointed Wardill stood down from the MCC committee after eleven years in September, although his send-off was a warm one with the parting gesture that the vice-president's chair 'to which his services on behalf of the club eminently entitled him' was his should he seek it.

Then, fleetingly, in the winter of 1873, Wardill must have believed that his luck had turned. A newlywed Dr Grace abruptly agreed to tour, provided his honeymoon expenses were paid, and word that the world's greatest cricketing attraction would be in Australia in December fired colonial passion. The visit would be another red-letter date in Australian cricket history. But Wardill would not witness it.

Auditors who had been studying Victoria Sugar's books had discovered on Friday 15 August a discrepancy of £200, with the finger of suspicion pointing to Wardill. Not only was Wardill a bad gambler, he also proved in his hour of adversity a lousy liar: he had the entirety of Saturday to escape, but return to Mona Place overwhelmed his first thought of suicide. 'I had determined to drown myself,' he wrote that evening. 'But when I went home to dinner the sight of my wife and child sent me to my bedroom and I prayed to God the first time in many years … God knows that I would do anything in the way of restitution if I had it, but my furniture is all that I have … It is a dreadful crime that I have committed and I would undergo anything if only my innocent relatives would not suffer. I think it better that I should make away with myself but, if I do, I have been taught that I shall be consigned to everlasting punishment and I dare not.'

Wardill set himself to make a clean breast instead and rose on Sunday to visit the home of company secretary Joseph Robertson, confessing there that his depredations had run to £7000 over five haunted, guilt-ridden years. He agreed that he would have to be charged and asked only that he be permitted to explain the affair to his wife and one-year-old son. But at home his courage finally buckled. Leaving his watch and a pathetic pile of silver, he scrawled a final note, still among the depositions on file at the Victorian Public Records Office: 'I have gone to the Yarra. It is best for all.' A search of the house discovered him gone and the Yarra ferryman who heard Wardill's splash that evening did not even think to report it.

If Wardill's death spared him the appalled glare of the people of Melbourne, he was right about the damnation. *The Herald* allocated him a seven-deck headline:

MANIA FOR EMBEZZLEMENT

DEFALCATIONS OF AN ACCOUNTANT

A CRICKET CELEBRITY EMBEZZLES OVER £7000

FULL EXTENT OF HIS DEFALCATIONS NOT YET KNOWN

HE CONFESSES HIS GUILT

LEAVES HIS JEWELLERY BEHIND AND DISAPPEARS

A WARRANT IS ISSUED, BUT HE IS SUPPOSED TO HAVE

COMMITTED SUICIDE

Rumour flourished over the next seventeen days. Eyewitnesses claimed that they had seen him board a ship bound for the United States; others swore he had headed for the bush. Finally, Wardill's body was discovered by the son of the ferryman, wakening curiosity again. Hundreds of voyeurs thronged the doors of the City Morgue, hopeful of a viewing; they were turned away.

'The worst feature of the case is that he married while all this was hanging over him,' ranted *The Sketcher*, 'and so has been the means of bringing an innocent family to shame ... Scandal has been busily engaged ... but it would serve no good purpose to give further publicity to the tales which are flying about. It is better that the man and his misdoings should be "to dumb forgiveness a prey".' Gambling, briefly, was in the dock. *The Herald* demanded that the 'jovial and open-hearted gamesters' who had profited by Wardill's losses be named: 'We trust they are satisfied with their handiwork and can look back with satisfaction to the merry nights they spent in helping a weak and foolish young man to his grave.' An anonymous letter-writer to *The Herald*, who seemed uncommonly well-informed, scoffed: 'It is asserted that he was led away by his gaming companions ... in my opinion Wardill was not a man to be led by anyone – he was a leader in everything and those who knew him best, who were daily in his company, I challenge to deny this.' *The Australasian* reminded cricketers of their 'enormous debt' to Wardill 'however great his crime', while the day after an inquest returned a verdict of 'temporary insanity', the MCC annual meeting diligently minuted his death: 'Mr Wardill, whose lamentable loss the committee bitterly deplore, stands fourth in the list of batsmen.'

*

Lamentation and deploring gave way, in time, to shame. Wardill was the first Australian to be entered in *Wisden Cricketers' Almanack*'s enduring section 'Births & Deaths of Cricketers'; his entry was, in the next edition, removed, perhaps as news of the circumstances of his death reached its editor.

Wardill's key role in MCC's evolution goes wholly unacknowledged in its official history. Jack Pollard's five-volume chronology of cricket in this country disposes of his inaugural century in barely a paragraph. Geoffrey Blainey wastes not a word on Wardill in his definitive exposition of the birth of Australian football, and Colden Harrison's self-congratulatory memoir of the code's early history makes no reference more detailed than to my 'good friend, the late Dick Wardill'. Wardill's great ally Tom Wills was destined for an eerily similar fate six years later, stabbing himself to death in May 1880; but in Martin Flanagan's celebrated tribute to Wills, *The Call*, Wardill does not appear either.

The reason may not be unconnected to Wardill's brother Ben, who was appointed secretary of the MCC in April 1879. Major Ben – the rank was honorary, betokening his enthusiasm as a volunteer in a local militia – served thirty-one years. Busy, businesslike, with a neatly trimmed Mark Twain moustache, he became what *The Australasian Star* called 'an international institution': perhaps the best-known cricket administrator of his day. But to do so required him placing a distance between himself and the memory of his unfortunate elder. When a Wardill Stand was erected at the MCG in 1911, there was no doubt in whose honour it was christened.

What remains of Richard Wardill's great feat, remarkably, is Montgomery's cup: a relic of them both, the cricketer who fancied acting, the actor who fancied cricket, both destined for self-destruction.

Its engraving is as sharp as the day it was presented, and it has a careful custodian in Dick's delightful great grandson, David Richard. The lack of a more permanent memorial might not be appropriate where Dick Wardill is concerned. He never looked back, nor very far forward – which might be a family trait. David Richard relates how his own grandson took a fancy to another small relic of Dick's – a mounted bowling award – and knocked off all its gold lettering by cheerfully belting it round the backyard. Not an action out of keeping with the spirit of Dick Wardill when he was, to quote Shakespeare, and Walter Montgomery, 'a very riband in the cap of youth'.

The Independent Monthly December 1992

G.H.S. Trott
The Madness of King Harry

CRICKETERS GREAT AND HUMBLE meet one opponent equally. Sooner
or later, all defer to time. Usually the end comes stealthily: muscles
stiffen, reflexes slow, eyes dim. There is time for grace, for goodwill,
and a formal farewell: Steve Waugh's imminent departure shapes as
a tearjerker. In some cases, though, it is abrupt, unceremonious,
unkind; so sudden that it's as though the individual was never there
to begin with.

One hundred and five years ago, Harry Trott led Australia to its
first win in a five-Test Ashes series. At thirty-one, he was in his athletic prime, and by common consent the world's wisest cricket captain.
In his watch had emerged what historian Bill Mandle has called 'the
unfilial yearning to thrash the mother country', even as the constitutional undergirding of an Australian commonwealth was being
erected. But at the peak of his fame, Harry Trott disappeared, never to
play let alone captain another Test. And though he reappeared as a
man, it was as a marginal figure, and a source of stifled embarrassment. For – didn't you know? – poor Harry went mad.

Trott was born on 5 August 1866, third of eight children. Father
Adolphus, an accountant, was scorer for the South Melbourne
Cricket Club, whose batting and bowling averages eighteen-year-old
Harry topped in his inaugural season. He was marked out from the
first by uncommon temperament, batting with easy grace, giving his
leg-breaks an optimistic loop. As befitted one destined to spend his
whole working life in the Post Office as a postman and mail-sorter,
he met everyone alike. On one occasion he was introduced to the
Prince of Wales, who after a long and convivial conversation con-
ferred on him a royal cigar. Later he was asked what he had done

with it: the fashion was for preserving anything of royal provenance as a keepsake. Trott, though, simply looked puzzled. 'I smoked it,' he said.

In his equanimity, Trott was a rarity. Australian teams of the time were notoriously combustible, riven by intercolonial jealousies. Trott's first skipper Percy McDonnell survived a 'muffled mutiny' against his leadership because of his over-reliance on New South Welshmen; his second, Jack Blackham, was weak, suggestible, and felt to be the cat's paw of other Victorians; his third, George Giffen, asserted South Australian primacy simply by bowling himself interminably. Discipline was lamentable, with Australia's 1893 tour of England especially unruly. 'It was impossible to keep some of them straight,' complained their manager. 'One of them was altogether useless because of his drinking propensities ... Some were in the habit of holding receptions in their rooms and would not go to bed until all hours.'

Appointed to lead Australia's next trip to England in 1896, Harry Trott changed everything. He knew no favourites, was never remembered to quarrel with anyone, and showed a pioneering flair for tactics. Rigid field settings and pre-determined bowling changes were standard operating procedure at the time; Trott set the trend of configuring fielders and employing bowlers with particular batsmen and match situations in mind, and rotating his attack to keep its members fresh. 'His bowlers felt that he understood the gruelling nature of their work,' said *The Referee*, 'and that they had his sympathy in the grimmest of battles.' To some, this sympathy seemed uncanny. Trott's star bowler Hugh Trumble had days, he confessed, when his usual sting and snap were missing; Trott could sense such occasions within a few balls, and would whip him off. Sydney batsman Frank Iredale, meanwhile, was gifted but highly strung, a teetotaller. 'Look here, Noss, what you need is a tonic,' Trott counselled. 'I'll mix you one.' Iredale made more centuries than any other batsman on tour, fortified by what Trott later admitted was brandy and soda.

In the end, the Australians were pipped 1–2 in the series, the defeats being narrow, their victory stirring. Opening the bowling on a whim at Manchester, Trott had England's stars W.G. Grace and Drewy Stoddart stumped in his first two overs: a decisive break-through. His leadership was then seen to even better advantage when England next visited Australia, and was overwhelmed 4–1. 'It didn't seem to matter to Mr Trott whom he put on,' lamented Stoddart, 'for each change ended in a wicket.' *Wisden* thought him 'incomparably the best captain the Australians have ever had'; the Anglo-Indian batting guru Ranjitsinhji concluded that he was 'without a superior today anywhere'. The Melbourne Test of January 1898 not only drew many of the delegates from the marathon Constitutional Conven-tion smoothing the path to Federation, it wholly distracted the public. As John Hirst says in *Sentimental Nation*: 'The Australian cricketers were better than the English. Who cared whether Aust-ralian judges were up to the mark?' Victory, though, was also felt a vindication of the cause. *The Bulletin* exulted: 'This ruthless rout of English cricket will do – and has done – more to enhance the cause of Australian nationality than could be achieved by miles of erudite essays and impassioned appeal.' Trott joined the Albert Park branch of the Australian Natives' Association, and his double-fronted weath-erboard in Phillipson Street became a place of pilgrimage. Trott's employers, when some complained about the leave lavished on their most famous functionary, explained: 'Harry Trott is a national institution.'

The precipitous decline of that institution, veiled in euphemism for more than a century, began on 8 August 1898. Visiting mother Mary in Doncaster, Harry Trott collapsed; in the words of *The Argus*, he 'fell suddenly into a fit', losing consciousness. With wife Violet on the train home, he suffered another violent paroxysm; despite a doctor's presence, he was convulsed a third time at 10 p.m. For four weeks, Trott lapsed in and out of consciousness, unable to work, barely able to communicate. Public anxiety was acute. When *The Australasian* stated it was 'an absolute necessity that he should have a trip into the country where he will have the benefit of ... a much-needed rest', supporters subscribed more than £400. Trott left Melbourne on 10 September with Test team-mate Harry Graham and a nurse, bound for Woodend; *The Woodend Star* reported his intention of 'recovering from a recent severe attack of illness'. But he was unchanged by his fortnight at 'Bon Air', a private retreat, and many were shocked by his appearance at the MCG on 15 October bearing 'unmistakable traces of the terrible sufferings he had undergone during his severe illness'. He was tormented by insomnia, loss of memory and an enfeebling apathy. Hope he would recuperate in time to lead Australia to England to defend the Ashes gradually evaporated. On 3 May 1899, a green and gold flag fluttered over London's Inns of Court Hotel advertising the arrival of Australian cricketers under the captaincy of Joe Darling; that same day, a world away, Darling's predecessor entered Kew Asylum.

*

Opened in December 1871, Kew Asylum served as a mental hospital for 117 years. In that time, observes Dr Cheryl Day in her thesis *Magnificence, Misery and Madness*, 'any resident or long-term visitor to Melbourne understood what was meant by the phrase "you should be sent to Kew"'. The building, however, was representative of a brief moment in psychiatric care when giant asylums seemed the way of the future. Kew echoed the theories of British alienist John Conolly, superintendent of asylums at Colney Hatch and Hanwell in England, who had recommended first and foremost 'a healthy site, freely admitting light and air … on a gentle eminence in a fertile and agreeable country'. It was believed that Kew's position – set on 170 hectares, thirty metres above river level, eight kilometres from the city – would 'conduce to the happiness and comfort of patients and facilitate their recovery'. Built in the shape of an E enclosing two courtyards surrounded by roofed walkways, with a facade and steep mansard roof in the French Second Empire style, it was the most ambitious and expensive such project in Australian history.

Yet within five years mental health meliorists had changed their minds. A government inquiry concluded: 'It has been well said that a magnificent asylum for the insane means, and must ever mean, the crowding of cases together which ought to be kept apart.' A royal commission eight years later investigating asylums at Kew, Yarra Bend, Ararat and Beechworth condemned them all: 'It is hard to say why the dwellings of lunatics should cost such extravagant sums as were lavished on the edifices in question.' The commission concluded, however, that the sunk cost justified keeping asylums open, and that the government might as well exploit their one advantage: economy. Policy trends reversed altogether: 'In dealing with a large number of pauper lunatics ... their actual wants and powers of appreciation should be regarded. Palatial residences, equipped with Chippendale furniture, would not make them happier, or minister to the wants of minds diseased. The majority of these people belong to the poorer classes and if they were supplied with the most elegant surroundings, they would be unable, even in their most lucid moments, to appreciate the refinements of better life.' Kew Asylum became Melbourne's Potemkin village of the insane. Boosters saw its external splendour as an index of wealth and sophistication: 'Our prisons and madhouses prove our civilisation to be equal to that of any of the older cities of Europe.' Inside, the great madhouse was growing madder. Designed for 600, it seldom accommodated fewer than 1000, ranging from eccentrics and the elderly to drunks and criminals – including several convicted murderers. For these, there were never more than three medical officers.

Not that Kew did not discomfit the city surrounding it. Melburnians were troubled from time to time that their colony seemed somewhat battier than it should be, the proportions of the population institutionalised in Victoria being far greater than in England or New South Wales. One noted alienist Dr John Springthorpe explained this by recourse to history: 'You may call it our fevered past – the time of the goldfields – the distinct nervous tendency inherited from those times, the excited natures that came out and which have been transmitted to their descendants.' Reforms were mooted. A new *Lunacy Act* in January 1889 foresaw the closure of both Kew and Yarra Bend, and a more rigorous admissions policy involving a receiving house and an inebriates retreat – plans stymied quickly by the 1890s depression.

The asylum mimicked some rituals of ordinary life. There was a cricket green, a tennis court and a billiard room, though mainly used by attendants. Patients were also expected to work; it wouldn't make them free, but might keep them calm. Kew's first superintendent had described 'employment' as 'the chief means of curing insanity of all kinds', and patients by the turn of the century followed a clockwork routine from 6.30 a.m. A basic farm operated, not to mention workshops and washing houses, while ornamental and kitchen gardens were maintained. But drudgery tended to deaden rather than discipline. In January 1898, *The Argus* stormed: 'A distinguished specialist attending patients at Kew stated only the other day that the state of things at that asylum was now so bad that a person just mad enough to be put under restraint ran the risk of becoming much worse instead of receiving medical treatment likely to restore his senses.' This risk Harry Trott's friends were prepared to run.

*

The contents of Kew's calf-bound casebooks reveal a world rich, strange and occluded. Two cases admitted the same week as Harry Trott convey that world's extremes. One, an ageing bootmaker suffering 'mania', believed that the Salvation Army had paid for his murder and that his tongue would be cut off; he 'took possession of the WC' and covered himself in excrement in order to ward attendants off. The other, a 'thin, pale and anaemic-looking' mining student, was afflicted with a 'melancholia' associated with his 'having practised self-abuse for some time'; the examining doctor noted his pessimistic conviction 'that people are fated to go through certain experiences which it is hopeless to strive against'.

Harry Trott might have taken this counsel to heart. Standing in one place, showing no interest in surrounding events, he seemed lost within himself. 'Refuses to converse not appearing to be able to follow what is said to him,' summarised a doctor. 'Answers questions in monosyllables. Does not rouse up when subjects are spoken of that formerly he was keenly interested in – has a vacant dull expression.' After four days, Trott was given an enema; his 'bowels freely moved, a number of hard masses coming away'. His colon had no effect on his consciousness. Sent optimistically to the cricket green, he took 'no interest', batting and bowling 'in a mechanical, indifferent fashion'.

The official diagnosis was 'dementia', the official cause 'alcoholic', though both reflect medical convention rather than empirical evidence. Kew's superintendent Dr William Beattie-Smith believed that alcohol was 'the most common cause of insanity', and that 'all the pity in the world will never conquer a weak will, selfish desires and moral perversion'. Nor was Beattie-Smith's recommended cure – 'manual labour, cheerful society, liberal and wholesome diet' – ever likely to be efficacious. Baffled staff took one chance. Kew Asylum traditionally prohibited newspapers, convinced that isolation had 'a sedative effect' and removed 'any delusions which they [patients] may entertain'. After a few weeks, Trott was shown newspapers sent to him, probably Darling's doings in England. His 'heavy, dull expression' was unaltered. Trott's identity as sportsman, indeed, was steadily subsumed by his identity as inmate. Between 24 June and 14 December 1899, not a single case note was taken. One whom eighteen months earlier had been among Australia's most famous was now among its most anonymous.

A remarkable vignette survives of Trott at this time. Paul Farmer was a Collins Street physician who, for obscure reasons, was abducted by friends on 13 September 1899 and briefly incarcerated. In *Three Weeks in the Kew Lunatic Asylum*, a sixty-four-page pamphlet about his ordeal privately published the following year, Farmer profiled several fellow patients. 'Case VII' was almost certainly Trott: 'Here is a well-known cricketer, whom we once treated as a hero. But alas! Like everything else, times have changed, and he is almost forgotten. He stands about and is very reserved. Probably he has been badly treated, or thinks he has been, by some person or persons, and is one of the many disappointed men.' Trott's immiseration, in fact, suggests what would now probably be identified as a classic depressive psychosis and treated either pharmacologically or with electro-convulsive therapy; Kew's medical staff at the time, however, were merely doctors, who observed the mind but treated only the body. Nothing, as

Farmer noted, prevented deeper loneliness: 'Surely such a man as this might be tended by someone more closely related to him and given his only chance of being restored to health. If he has gone mad in the cause of cricket, then the cricketers should see that he gets a few of life's comforts and look him up occasionally. For though that man may appear indifferent to such attentions, he is not so.' Farmer's conviction about Trott's isolation makes sense. The Melbourne Cricket Club's secretary Ben Wardill visited at the end of November, but left unrecognised; no others from the sporting community seem to have made the journey. Not, at least, until Trott trod tentatively back into the sporting community.

On 10 February 1900, a Kew Asylum XI hosted a club called North Melbourne Rovers. Not surprisingly, the quondam Test captain appeared, albeit with a man to run for him, on account of his leaden gait. To universal astonishment, Trott began swinging his bat in mighty arcs: his 98 in forty minutes contained twenty fours and a six. A doctor examining him shortly after cautioned against overoptimism: 'Considerable mental dullness still exists. Memory still impaired. Reads papers sometimes but does not remember the contents much.' But when Wardill took Test keeper Alf Johns to Kew a fortnight later, Trott piped perkily: 'Hello Alf! How are you?'

'Dr Smith says Harry played really well,' Wardill reported in a letter to Trott's South Melbourne colleague Bob McLeod. 'There was no soft stuff set up for him to hit, and he bowled and caught well. He was more talkative and brighter altogether.' Muscle memory reactivated other recollections; he played a string of forceful innings against other visitors, and in April secured 6/30 including a hat-trick against a team from the Commercial Travellers Association. Trott was moved to be presented in token of his feat with a hat. He was famously fond of them; a team-mate once described him as 'the only man I have seen who, in the nude, had to have a hat on his head'. The doctor who visited the following week thought his 'improvement marked'. After 400 days, Trott's file was inscribed: 'Discharged. Recovered.' But not, perhaps, rehabilitated.

Though only thirty-three, Trott was never again a conspicuous cricketing figure. A transfer to Bendigo Post Office limited him to half a dozen first-class games in seven years. There was little diminution in his abilities: returning to South Melbourne in 1910–11, he again led batting and bowling averages. But the circumstances of his Test career's end were henceforward unmentioned. Trott himself alluded to them only once, during a December 1913 skirmish with the Victorian Cricket Association's Mat Ellis in the letter pages of *The Argus*. 'Mr Ellis asks me why [John] Giller did not go to England when he was at his zenith as a cricketer,' wrote Trott. 'This was in 1899 and Mr Ellis, unless he is still suffering from bad memory, knows that months before that team was chosen I was laid up with a very serious illness and had no say in selection.' Ellis was probably mortified. When Trott died in November 1917, it was he who moved that the VCA raise money for a suitable tombstone. The settings of much of Trott's life have been erased by time – even the green on which he played his resurrecting innings has gone, occupied by a residential development near the long-dormant asylum. But Trott's memorial, a marble column on a granite base in Brighton Cemetery, survives.

The June 1919 dedication ceremony was a curious affair. The war was not long concluded, and the address by VCA president Donald McKinnon, also Director-General of Recruitment, was replete with martial allusions: 'Our greatest heroes were those 60,000 men who had died to keep the country free and the men who had fought with them. Harry Trott was a hero because he had excelled in a great sport.' In recognition of Harry Trott's inner battle, of course, no laurels were bestowed.

The Bulletin Summer Special 2003–2004

The Second XI

H.V. Hordern
Googlies

HERBERT VIVIAN HORDERN: IT IS THE KIND of name that should have been attached to a foreign minister, an admiral or even a clergyman. Fortunately, cricketers did better with his nickname. 'Ranji', they called him, because of his dark complexion, with the Indian nob turned English batting gent in mind – and like the original Ranji, he was a law unto himself.

For his historic significance, one must cast back to Sydney on 3 March 1904, when the Englishman B.J.T. Bosanquet had his first big day out with the off break he'd perfected that looked to all the world like a leg break. Six Australians, peering sightlessly down the pitch, succumbed to him for 51; England reclaimed the Ashes.

Australia has a chequered history with introduced species, and 100 years ago the googly loomed like the rabbit and the prickly pear. But sitting in the crowd that afternoon trying to discern its deceptive geometry, twenty-one-year-old grade leg spinner Hordern ached to master this weird new science.

It is sometimes said that Hordern stole the googly, yet what he really did was re-imagine it. There was no spin cam or super slo-mo to help him, of course; he had no coach to consult or book to peruse. He had to intuit himself from first principles how the reversing of the wrist reversed the effect by tilting the axis of spin. It took eighteen months.

As the 1905–06 season commenced, there came what must rank as one of Australian cricket's 'eureka' moments. Hordern invited a New South Wales team-mate Alick Mackenzie to the Sydney Cricket Ground on the pretext of 'trying something', and timidly threw up three googlies. Two full tosses were 'promptly hit from one end of the ground to the other'; the third pitched, arrowed in, and beat the bat. Asked to explain his misjudgement, Mackenzie was puzzled: 'I don't know. I just missed it.' Hordern shortly became the first Australian to unleash the googly in a first-class match: his effect was immediate.

The ball is cricket's oldest accoutrement; the challenge of making it 'do' something predates even the jelly baby. A century ago, the googly was like reverse swing today: an invention seeming to turn cricket on its head. Some obscurantists thought it illegal; Bosanquet himself insisted that it was 'merely immoral'.

Truly exquisite in Hordern's story, however, is what happened next. Like a veritable human googly, he did the opposite of what you'd expect, choosing to study dentistry at the University of Pennsylvania. While South Africa gave rise to no fewer than four expert exponents of the googly – Vogler, Faulkner, Schwarz and White – the Australian who bowled it best was representing the Gentlemen of Philadelphia, visiting the Caribbean once, England and Ireland twice, Canada thrice. An obliging university team-mate filled his locker with Bulli soil to forestall homesickness; it seems to have been as much Australia as he needed.

Hordern's 'have-googly-will-travel' lifestyle lasted four years. He was known in Toronto as purveyor of the 'boomerang ball', and it came back wickedly. 'My heavens!' exclaimed MCC's Ted Wynyard as he was bowled shouldering arms. 'A googly!' West Indians simply stared in wonder. While touring Jamaica in 1908, Hordern awoke one morning to find a group of hotel workers examining his hand for a sixth finger.

This we know because Hordern wove it all into a mighty yarn, and if a more enchanting Australian cricket book exists than his 1932 memoirs I don't know it. In this 'evening's chat by an Old Buffer to some Unknown Friend', Hordern savours his country, county and club days, including representing Canton (against Mandelong with its eight fearful Kellys), London County (with W.G. Grace, who 'ran occasionally for himself, and never for his partner'), and Scotland's Leith Caledonian CC (where everyone played under an alias: Hordern's was 'Dr Fiddlesticks').

Hordern's first-class figures are impressive by any standard – 228 wickets at 16.4 – but you'd never guess from his stories. Few memoirs have described in such loving and hilarious detail the experience of failure: ducks, dropped catches and nones-for-plenty. A chapter entitled 'A World's Record' is devoted to his *three* noughts in a game at North Sydney ('Now bring out your Don Bradmans!'). He recalls how E.L. Waddy once hit him so far so often in a game at Maitland that a fielder who'd retrieved one of the sixes did not return, taking tea with some girls he met instead. He reflects on being recognised in retirement: 'I frequently get, "Oh, I remember you well, it must be years and years ago now, bowling in such and such a match", and then follows a detailed account of how some forgotten batting giant of other times trounced my bowling all over the place.'

Then on the last dozen pages of more than 200, almost as an addendum, Hordern relates how he arrived back in Australia late in 1910, and thought he might have a bit of a crack at this Test stuff, which was 'a rather serious matter' but 'really splendid' – he omits to mention, of course, his 46 wickets at 23 in seven appearances for Australia. To the very end, in fact, Hordern's book is in almost every respect the opposite of what you'd expect. But perhaps that is why he entitled it *Googlies*.

The Age January 2004

Reg Duff
The Second

WHILE CRICKET'S GREAT PARTNERSHIPS trip readily off the tongue, they are seldom truly equal. Somehow, all contain an order of precedence. It will always be Lillee and Thomson; it will never be Sutcliffe and Hobbs. For some, this means a steady eclipse, and eventually a historical redundance. With a name that always suggests a character from Baroness Orczy, Victor Trumper still commands a place among cricket's most hallowed reputations; but who today could tell you aught of his opening partner Reg Duff?

It's almost thirty years since I chanced on Duff's name in one of my first cricket books – I expect it has been the first for many readers – *Cricket the Australian Way*. Commenting on Bill Woodfull and Bill Ponsford in a chapter on the masters of the past, Arthur Mailey mused that they were 'every bit as difficult to remove perhaps as Trumper and Duff'. While I knew the other three legends, Duff I did not recognise. Yet here he was shorn of his Christian name, as though it was entirely unnecessary to mention it.

This pricked my curiosity. Duff had, it turned out, been one of those great Australian selection 'roughies'. Twenty-three years old, and with only one century to show for nineteen first-class matches, he paid off like a veritable Peter Taylor when chosen for the Melbourne Test ushering in the year 1902, rising above a rain-ruined pitch to make 32 in Australia's first-innings' 112 at number seven, and 104 of its second-innings' 353 at number ten.

For much of the rest of his career, he would go in first with Trumper, a partnership notable for its *allegro* tempo: their 135 in seventy-eight minutes at Old Trafford later that year and 129 in eighty-eight minutes at Adelaide eighteen months later prefigured great Australian victories. He bookended his twenty-two Tests with another spirited century, 146, in his last appearance for Australia at the Oval in 1905.

In pictures, Duff always cuts a fine figure: dark, unsmiling, slim at the waist, broad in the shoulders, full of compressed power, but also with a slightly dandified air from his moustache, which like the Kaiser he curled upward at the ends. He was most picturesquely described by the English captain and polymath C.B. Fry: 'Reggie Duff, who had a face like a good-looking brown trout, and was full of Australian sunshine, was an entertaining batsman of excellent class.' According to Fry, Duff was also the first Australian batsman to drive routinely on the up – what Fry calls the 'uncommon stroke' of 'the high drive'.

Duff certainly seems to have lost little by comparison with Trumper in their collaborations for New South Wales. In December 1902, when they added 102 for the first wicket against Victoria in barely an hour, Duff proceeded to a personal 102 out of 198. The following month they amassed opening partnerships of 298 in 133 minutes against South Australia and 267 in 137 minutes against Victoria. When they were the first Australian opening pair to break three figures in both innings of a first-class match in January 1904, their stands of 113 and 119 lasted less than an hour each. Yet, while Trumper's fame endured, Duff's was fleeting. Most cricket tragics can tell you that Trumper was bowled by the first googly released from captivity in Australia, by B.J.T. Bosanquet, at Sydney in March 1903; only a rare one can cite Duff's feat of 194 in four and a half hours in the same match.

Why? *Wisden* remarked that Duff 'should have had a longer career'; *The Referee* commented that 'he did not take that care of his health necessary for one wishing to live a normal life in years and in vigour'. The reader of journalism from the period soon learns to identify such remarks as euphemisms for the curse of alcohol. Mailey's autobiography opens with a cameo of the author as a street urchin spotting Duff, 'one of my heroes', in Sydney's Chinatown, 'shabbily dressed and with his hair poking through his straw hat'. Despite the circumstances, Mailey recalls the thrill of the glimpse: 'Seen in the street that day he looked merely forlorn ... but the sight did not depress or disappoint me. That dapper little man had walked out with Trumper which was reason enough for him to remain one of my idols.'

Idol status, however, does not compel continued selection. In December 1907, Duff was twelfth man in the First Test against the visiting Englishmen. But he never again appeared for his state, and shrank from view so quickly that he was soon gone from grade ranks too. On the eve of Australia's next home Ashes rubber, former teammates gathered at Gore Hill for his funeral; Duff, under treatment for alcoholic poisoning in Royal North Shore Hospital, had suffered a heart attack aged thirty-three. Having been part of that sizeable category of cricketer, the stalwart somewhat overshadowed by his partner, Duff joined an even larger group, the figure somehow overlooked by history – perhaps originally through embarrassment meeting discretion, but eventually, simply through the passage of years.

The Age December 2003

Peggy Antonio
The Girl Grimmett

THE VOICE ON THE OTHER END OF THE PHONE is cautious, but immediately warm. 'Peggy Antonio?' it says. 'Yes I am. Or I was. I've been Peggy Howard for almost fifty years.' The lady leg spinner of Australia's first women's cricket Test series in the 1930s? The 'Girl Grimmett' so brilliant that male critics debated the possibility of unisex Test matches? 'Well, I don't know about that,' comes the reply. 'That was a very long time ago, and I'm afraid I don't remember much.' All the trophies, souvenirs, memorabilia, it's explained, have long gone or been recycled. The mounted balls were unscrewed for the children to play with. The old Australian blazer made quite good dusters. But she's happy enough to talk.

Peggy Antonio always deplored the fuss about her, but her story has an irreducible media appeal. She slots into every disenfranchised group you can conceive of: an impoverished teenage girl of the Great Depression raised by a lone mother, one of six in a family mingling Chilean, Spanish, German, French and English blood. Her job: making boxes in a shoe factory. Short, too: five feet in sprigs. *And* a leg spinner. Such a good one, though, that at fifteen she was representing her state, at seventeen starring in the first women's Tests between Australia and England. At twenty, having been able to make Australia's first women's cricket tour only through the intervention of a gentleman patron, she was a celebrity in England.

Faithful to the tradition of her craft, too, she saved her most baffling delivery until last: after that 1937 trip she never bowled in competition again. But she is recalled, nonetheless. Last year, in an article marking a hundred years since a woman first slipped into its pages, *Wisden* invoked her above all as a women's cricketer who could

have crossed the gender gap. It recalled how Neville Cardus rhapsodised after seeing her: 'Suppose one day the greatest slow left-arm
bowler in England is a woman, will any male selection committee at
Lord's send her an invitation?'

'*Wisden?*' says Peggy. 'That's a record, isn't it?'

Nelson Raymond's Shoe Factory in Collingwood has long since
been erased by time, but its formation of a women's cricket team and
ground in 1931 stamped it as a pathfinder. Raymond's Club was one of
half a dozen when founded, but that number grew to forty in the next
eight years. Suburbs mingled with societies and dark satanic mills in
the burgeoning women's league, the YWCA and the Labor Social Club
mixing with the likes of the art needlework firm Semco and shirt-
maker Pelaco.

Top of the shorthand and typing honours at secretarial college
she might have been but, with 70,000 unemployed in Melbourne
alone, Peggy was happy to find a box bolthole at Raymond's: 'I went
for plenty of jobs and they always said, "We're really looking for
someone with experience." Now how was I to get experience?' As far
as Raymond's was concerned, however, fifteen-year-old Peggy already
had experience: leg spin. She'd learned it from boys in the streets
around her Port Melbourne home. While there had been no father to
foster her keenness – a Chilean docker of French and Spanish parents,
he had died when Peggy was fifteen months old – there had been an
uncle Jack. The family backyard evolved into a small-scale replica of
the MCG, on which Peggy directed over after over of bowling at a
single spot on the pitch. No spinning straight away. Just metronomic
pitch-and-toss until perfect accuracy was achieved.

Women's cricket matches at the pit of the Depression could draw
crowds of up to 5000, Raymond's reflected, so a works team might
in time be a shoe showpiece. It was a time when female athleticism
was finally being acknowledged, with the implicit sanction of even
Buckingham Palace: the Prince of Wales solemnly declared in 1934
that he 'saw no reason women should not wear shorts'. Raymond's
did become a heavy hitter in the competition, and the crowds did
blossom. 'Oh, it was marvellous,' Peggy beams. 'No-one could afford

to go anywhere so they'd just turn up at the local match. And the barracking! We had fights breaking out. Among the men, of course.'

One of those men would prove instrumental in Peggy's career. Eddie Conlon, a Melbourne club cricketer and something of a ne'er-do-well, was an encyclopedia of the game but a closed book in every other way. He became Peggy's mentor, teaching her a full repertoire, and forcing her to rehearse on the turf practice pitches at South Melbourne. The books before bedtime he prescribed were two instructional texts by Clarrie Grimmett. The finished Peggy could bowl leg spin and off spin, a rare mix, with a top spinner and a wrong 'un used frugally: quite a package. Too good, in fact, for most of her sorority. Although just seventeen, she was thought a banker for selection when the Australian Women's Cricket Council threw out a challenge to its rival in England in January 1934.

It wasn't an overgenerous offer. The AWCC at the time had 4s 7d in credit. Its invitation extended to hospitality and billets, but the English players would have to cover their own passages. Fortunately, they could. The visitors who stepped from the *Cathay* in Fremantle in November came from a country where the women's game was becoming a kind of polo for the ponyless. Their hosts came almost exclusively from working-class suburbs. When it was suggested that the women's Tests be played for their own set of ashes, the AWCC demurred: it was hard to justify the expenditure. But the council donned its glad rags and Betty Archdale's team was feted across the country, especially star batswoman Molly Hide.

There was a luncheon hosted by the Melbourne Cricket Club – which still brings a faint blush to Peggy: 'I had to stay at work at Raymond's as long as I could and, of course, the cable tram up Smith Street was late. They'd all sat down by the time I got there and I walked in: seventeen years old and everybody watching.' Peggy soon retrieved her poise, representing her state against the tourists at the MCG. Her 10/48 included a hoodwinked Hide. 'She was a flowing player, graceful, always casual, four, four, four,' Peggy recalls. 'But, do you know, I got her with an ordinary off break. I can still remember the look on her face. She didn't know a thing about it.'

Crowds grew to rival many Sheffield Shield crowds that season. Again they were mostly men, although their barracking never lapsed into coarseness. Even the Sydney Hill's celebrity spectator 'Yabba' Gascoigne was placated when he watched the English girls play New South Wales. Asked about his serenity, the progenitor of 'ave a go yer mug' and 'get a bag' said dismissively: 'The ladies are playing all right for me. This is cricket, this is … leave the girls alone!' Peggy became a crowd puller. Her twelve wickets in the series when she was only eighteen, including 6/49 in the last game at the MCG, were all top-order scalps: she became an adolescent antidote to a grey era. She was suitable even for publicity dates with Don Bradman at a Spencer Street hotel, although Peggy's main recollection of him is, despite her lack of inches, not having to look up very far. She never really liked his flint-hardness on the field either.

When England issued a reciprocal invitation for an Australian team in 1937, again, only hospitality could be offered and the £75 passage was well beyond her family's means: roughly a year's basic wage for a woman. 'My family didn't have a razoo,' Peggy says. 'Only one of my sisters was working and she was married. It was out of the question.' But cricket's grip on the nation was not to be underestimated. The country's men were winning the Test series in such exhilarating fashion that the injured survivors of a famous plane crash in Lamington National Park in February 1937 greeted their rescuers with: 'What's the Test score?' In Port Melbourne, raffles and a dinner dance were organised. Raymond's pitched in a set of travelling cases and a folding Brownie camera. Then, Peggy opened *The Herald* to the page regularly given to women's cricket and its correspondent Pat Jarrett. The 'Girl Grimmett' could go to England after all. An executive of McIlwraith McEacharn had pulled £100 from his own pocket to ensure Peggy's passage.

It was the start of a beautiful financial friendship. James McLeod had, he explained, known Peggy's father on the waterside, and requested only the honour of occasionally entertaining Peggy at his palatial Hawthorn flat. He would sponsor her throughout the tour, sending her a letter of credit and periodically alerting her by mail to his latest deposit. There was certainly none of *that*. 'He never made any … how shall I put it? He didn't ever push himself,' Peggy recollects. 'He didn't make me feel I was under an obligation. He had a friend, and he used to send his car round and pick me up from Raymond's and take me to dinner at this friend's flat.' Needless to say, there was a certain novelty to the sight of an urchin box-maker perched in a limousine in Collingwood streets during the Depression, and Peggy remembers especially the response of Clarendon rival Ruth Tucker. 'I was sitting in the back of the car when it pulled up next to hers at the lights,' she says. 'Ruth's eyes just popped out of her head.'

The regulations in force on the Australians' journey to England are worth quoting in full: 'No member shall drink, smoke, or gamble while on tour. No girl may be accompanied by her husband, a relation or a friend. Writing articles on cricket during a tour is strictly forbidden. While on board ship, no girl shall visit the top deck of the liner after dinner. Members of the team must retire to bed by 10 p.m. during the voyage. Members will do physical drill on deck at 7.15 a.m. daily except Sundays. The team will participate in all deck games.' Manager Olive Peatfield singled Peggy out for special attention when it came to enforcing her regime. 'I was the youngest in the team,' Peggy remembers. 'And I think she thought: "Port Melbourne background? She'll have to have a brick put on her head." But I wasn't the sort, even at that age, to put up with that. Oh, I didn't kick over the traces and make a nuisance of myself or anything like that. I just looked the other way.'

Peggy shone as soon as the matches began. Player after player departed baffled by her bag of tricks. Runs poured from her bat. She admits today there was a bit of con amid the conjuring: the wrong 'un always eluded her. 'I never perfected it,' she says. 'I'd just run up, go through the action, and hope for the best. It's an unnatural action and you have to pluck up an enormous amount of courage. There were times it ended up at slip, even in a Test once.' But Marjorie Pollard waxed lyrical of Peggy in her tour diary: 'If I see a cleverer spin bowler than Antonio I shall not only be surprised but genuinely disbelieving.' Pollard broadcast summaries of the Tests on BBC radio and also persuaded Movietone News to send a crew, so news of Peggy's 9/101 in the Northampton Test and 8/65 in the Test at Blackpool spread far beyond the boundary. An extraordinary crowd of 10,000 descended on Mitcham Green for the Australians' last engagement, against Surrey. 'More prams, dogs, small children, ice-cream men and bikes than I have seen anywhere,' diarised Pollard. 'As the crowd increased, so the boundaries decreased. Trams, buses, cars crashed by on all sides.' The crowd gave Peggy an exuberant farewell salute. They were not to know it amounted to her farewell from the game.

Quite simply, cricket had stopped being fun. 'In the end, I got sick of the grind,' she says. 'You had to win, win, win. It soured me. Whenever I played, I had to do this and I had to do that.' Pre-season practice brought matters to a head: 'Jokingly, they said: "Now, we want you to make 100 and take 10 wickets." But I wasn't in the mood for jokes. I said: "I think you'd better put the twelfth girl in." No hard feelings. No-one threw a bat, but that was that.' Peggy kept making boxes. Eddie Howard, who asked her for a glide at a dance in Prahran, regarded cricket with blank indifference. An Englishman abroad and a 'sort of undiscovered literary type', he was hoboing round Australia. When they married in 1943, he settled down; he became a fireman.

Cricket has become more serious since Peggy took her cap and sweater for the last time, and the only thing she shares with 'that silly Merv Hughes' is that he occasionally throws in a leggie. 'I barrack for anyone who's playing against Australia,' Peggy says. 'My family won't even talk to me about cricket anymore.' One daughter has managed to collect some of Peggy's detritus over the years, although the only bat in her Eltham home was cut from a paling for the amusement of her four children. Her mother managed to save the pocket of Peggy's Australian team blazer when the rest of the garment was chopped up for rags.

Peggy's a bit sorry about the blazer now. Perhaps she was a tad hasty. 'It was a bit silly,' she admits. 'It's so hard to find a nice blazer these days.'

Wisden Cricket Monthly March 1993

David Hookes
Living in the Seventies

THERE IS NO SIMPLER WAY of explaining the appeal of David Hookes, and the sensations of loss that follow his senseless death, than to describe him: an Australian cricketer of the seventies who remained one into the twenty-first century.

Not that he was a reactionary or relic; on the contrary, he was a savvy media performer and an innovative coach. But he retained throughout his career a breezy confidence that recalled an earlier, perhaps more innocent, certainly more intuitive, sporting era. For the generation who grew up when Hookes was the face of Australia's cricket future, his loss is more than a death; it is a *memento mori*. Given that disclaimers are these days ethically *de rigueur*, it should be revealed that the author's first adult bat was that peak of 1980s willow kitsch, the Gray-Nicolls double scoop, signed by David Hookes.

Hookes' emergence is still one of Australian cricket's fairytales – how, barely twenty-one, he harvested five Sheffield Shield centuries from six innings. Personally, I still recall the headline in Australia's *Cricketer* magazine: 'Heck! It's Hookes!' A blonde, broad-shouldered, loose-limbed left-hander, he looked like he'd been sent down by central casting to play the prince to Snow White.

In this era of academies, structured careers and ordered change, Hookes might have had to prove his consistency over a longer course; in his day, perhaps more amenable to raw talent, he was at once slotted into Australia's middle order for March 1977's Centenary Test. Australian cricket was then on the crest of a five-year wave, set in motion by the Chappell brothers, sustained by the likes of Dennis Lillee, Jeff Thomson, Rod Marsh and Doug Walters, with supporters increasingly demonstrative of their allegiances. But it was more than simply a time in which Australia's Test team was strong; it was one in which telegenia counted. Hookes' record as an international batsman is modest, but he happened to feature in two of televised cricket's great moments: that Centenary Test, in which he struck five unforgettable fours from consecutive deliveries by England's Tony Greig, and the Second Supertest of Kerry Packer's World Series Cricket later that year, when his daring assault on a West Indian pace barrage was abridged, sickeningly, by a broken jaw.

Hookes had been chased by the breakaway World Series Cricket for his profile as well as his cricket potential, and his maiming by Andy Roberts paid back the whole investment. Until that moment, WSC had looked suspiciously like a thrown-together entertainment package; Hookes' injury, captured in graphic detail by the new all-seeing eye of Channel Nine's cameras, impressed on every viewer the contest's intensity.

Hookes did sterling deeds thereafter. In one free-swinging home summer, 1982–83, he gathered 1424 runs at 64, then topped it off with his only Test century, an unbeaten 143 in Australia's inaugural Test against Sri Lanka at Kandy. Yet those cameos in 1977 – one in the traditional game, one in the new-fangled affair that arose to challenge it – defined him thereafter.

They also define a hinge moment in cricket history, about which a certain ambivalence remains in even the most ardent sports consumer. Professionalism: wonderful, isn't it? Money: they deserve it, don't they? Mind you, the players these days ... they're a bit boring, aren't they? While part of the first generation of cricketers who looked on the game as a livelihood, Hookes retained his spontaneity. When he took on the role of hosting 3AW's *Sports Today* in September 1995, he very clearly held no brief for anyone. In a media community so anxious to ingratiate itself with athletes, and athletes so anxious to say nothing that can be misinterpreted that they are inclined to say nothing worth interpreting, Hookes reminded me of Dorothy Parker's reply when told that she was 'outspoken': 'By whom?'

When Hookes added the job of Victoria's state coach in May 2002 to his media duties, it was with typical brazenness. It is an irony that he should have met with his misadventure while indulging in one of the old-fashioned customs he had successfully inculcated in his charges: the post-match drink with comrades and opponents.

There is a temptation to contrast the grief for Hookes with the last significant mourning period in Australian cricket: that for Sir Donald Bradman. Bradman's was, of course, infinitely the greater reputation. Yet the marking of his passing involved a significant degree of, to borrow Douglas Coupland's phrase, 'legislated nostalgia'. We were invited, and in some instances instructed, to lament the passing of a man said to embody an era very few could remember, and which had in any case long predeceased him. The sadness in Hookes' death springs from him being part of a stronger collective memory. In the cruellest of circumstances, the youngest player in the Centenary Test has become the first to die. The seventies are still relived in hearts and minds, but a personification of them has gone.

The Bulletin February 2004

Mohinder Amarnath
Jimmy

TO PLAY SIXTY-NINE TESTS for one's country is an enormous honour. To miss sixty-four in that same period is to have endured some of the gravest disappointments cricket can inflict. The record of India's Mohinder Amarnath is unlike any other; it strangely befits a player whose career was of almost unique extremes.

He was known, universally, and in appreciation of his affable and equable temperament, as 'Jimmy'. If he did not loom among countrymen like contemporaries Gavaskar, Viswanath and Kapil Dev, he was a cricketer's cricketer: contributors to a 1996 book of tributes, *Grit & Grace*, laud his decency (Viv Richards calls him 'one of the nicest men to have ever played the game') and his determination ('Concede didn't seem to be in his vocabulary,' says David Boon).

He needed both. He was the last chosen in every reconstitution of India's Test team, the first excluded in every purge. At his peak just over twenty years ago, he was freely described as the world's best batsman; he then suffered a loss of form so abject and ruinous that it seemed almost ghoulish to look.

Much of Jimmy's career is explained by his father, Lala, one of India's most eminent batsmen and eristic personalities. A fearsome martinet, Lala forbade his three boys from playing any sport but cricket, prohibited their using anything softer than a cricket ball, and orchestrated games in his garden where pots stood in for fielders in order to develop skills of placement. In Indian cricket, meanwhile, the name Amarnath invited reverence or revulsion: the sins of the father were regularly to be visited on the son.

Jimmy made himself a still greater target. He batted, as his father demanded, with a flagrant disregard for danger. In an era replete with fast bowling and unrestricted in use of the bouncer, he never stopped hooking – despite many incentives to do so. He suffered a hairline fracture of the skull from Richard Hadlee, was knocked unconscious by Imran Khan, had teeth knocked out by Malcolm Marshall and was hit in the jaw so painfully by Jeff Thomson in Perth that he could only eat ice-cream for lunch. 'What separated Jimmy from the others,' Michael Holding said of him, 'was his great ability to withstand pain … A fast bowler knows when a batsman is in pain. But Jimmy would stand up and continue.'

It was as a combatant of fast bowling that Jimmy enjoyed his finest hour. He spent three years by the wayside after being maimed by Hadlee in 1979, but re-emerged with a helmet, a two-eyed stance and 1182 runs at 70 including five hundreds in consecutive away series against Pakistan (Imran and Sarfraz) and the West Indies (Holding, Marshall, Garner) during the first half of 1983.

Tall, slim, apparently impervious to fear, he advertised his resolution with a red handkerchief that always protruded from his pocket – à la Steve Waugh. I have some grainy video footage of Amarnath batting in the West Indies that is nonetheless absolutely stirring, the crowd noise drowning the commentators' voices as he hooks and pulls off the front foot. When he had to retire for stitches in a head wound at Bridgetown, he washed the bloodstains from his shirt while waiting to resume, and on doing so hooked his first ball from Holding for six. When India won the 1983 World Cup, he was man of the match in both the semi-final and final.

Then, a few months later in India, came a reversal of fortune as complete as any in history. Jimmy made 11 in two hits against Pakistan, and 1 run in six tortured innings against the West Indies. I followed international cricket closely in those days, had been thrilled by Amarnath's comeback, and can still recall my incredulity as the daily papers reported his steady accretion of failures – twenty years later, it still tweaks my curiosity like few sport stories.

What happened? Who knows? Yet every cricketer has had an experience like Amarnath's, that sensation of being the butt of a giant joke. Reality television has recently wakened viewers to the compelling nature of humiliation. Lovers of cricket have subtly acknowledged this for years – their game contains greater scope for humiliation than perhaps any other, in the defeated trudge of the batsman bowled for nought, the wounded air of the bowler hit for six, the shame of the errant fielder.

Here, again, Jimmy was the cricketer's cricketer. Having scaled Olympian heights, he shrank to the level of the most inept park bumbler. Yet there was, remarkably, another comeback in him; a year later, having recalibrated his technique again on more orthodox lines, he saved India with a gritty hundred in Lahore, followed it with three more, and was not displaced until the advent of Sachin Tendulkar – a player to whom, for all his greatness, I find it harder to warm. Tendulkar makes cricket seem inevitable; Jimmy Amarnath's career reminded you instead of its true evitability.

The Age November 2003

Tony Greig
Smiling, Damn'd Villain

CRICKET HISTORY IS RICH IN HEROES. Far fewer are those branded villains. The generation who've grown up listening to Tony Greig's Channel Nine sales schtick know little of the wrath he courted in March 1977 by selling himself.

When Greig learned of Kerry Packer's plan to create a professional cricket circuit during their fateful meeting in Sydney, he wrote himself a list of six pros and four cons. Pros included long-term financial security, cons the likelihood that he would, at least in the short term, be a sporting pariah. Correct about the pros, he perhaps underestimated the cons: today he is barely remembered as a cricketer. It isn't even clear to whose cricket history he belongs: South Africa, where he was born, England, whose team he led in fourteen Tests, or Australia, where he settled.

Greig's career is probably not simply unique but unrepeatable. While regarded as the first of the flag-of-convenience foreigners to populate English cricket, he was probably more 'British', insofar as such measures are meaningful, than many who followed: his grandfather owned a Bathgate department store; his father had been a Bomber Command squadron leader.

Greig, moreover, arrived in England in April 1966 unheralded, having played only one first-class match in the Currie Cup's 'B' section. Had his coach at Queenstown's Queen's College not been the Sussex pro Mike Buss, and had he not been ruled unfit for South African national service because of his epilepsy, Greig might have gone little further with cricket. As it was, he rapidly made his mark, with 1299 runs and 67 wickets in his first full season aged twenty. When the boom was lowered on South Africa in world sport, Greig found himself on the right side as much by accident as by design.

Greig passed 1000 runs in his fourteenth Test, took 5/24 against India at Calcutta in January 1973 with medium pace, then 13/156 against the West Indies at Port-of-Spain in March 1974 with off spin. According to Alan Knott, Greig was for a time 'the best off spinner I have seen in Test cricket'. 'When you batted against him in the nets you didn't know how you'd survive,' Knott recalled. 'He got bounce and really spun the ball, giving you the feeling that you were always going to be caught by a close fielder.'

His Test record – 3599 runs at 40.4, 141 wickets at 32.2 and 87 catches – puts him in the front rank. His commanding physical presence – blonde, blue-eyed, 200 centimetres– made him stand out all the more. Writers tended to lose perspective in describing him: *Wisden* sounded eerily like a Nazi eugenicist when it described him as 'the Nordic superman in the flesh'.

He came to the crease swinging his bat in uninhibited arcs, baited rivals as the silliest of silly points, amused and antagonised crowds with equal ease. And if he didn't devise them, Greig is closely identified with three features of modern batting now taken for granted: the stand-to-attention stance, which he adopted in India to counteract the bounce of their spinners; the slice over slips, by which he turned the pace of Jeff Thomson and Dennis Lillee to his advantage; and the helmet, whose 'motorcycle' antecedent he was among the first to wear.

So what happened? Of all the Packer signatories, Greig had most to lose and least to gain; he not only sacrificed the captaincy of his adopted country, but what shaped as a record-breaking benefit season at Sussex. Yet Greig did more than merely sign; he converted. Reasoning that 'cricketers were quite likely to be put on their guard by a businessman offering vast sums of money', he volunteered to recruit the next fifteen foreign players himself. When it emerged that he had acted as a WSC agent, apologists like Knott celebrated Greig as having done 'more for the professional cricketer than any other man'; but for most in England, Greig had less in common with Lord Hawke than with Lord Haw-Haw.

Greig had always been ambivalent about England's captaincy. After the tabloid torments of predecessor Mike Denness, he'd inwardly promised to make the best of his peak cricket years then 'get out before I could be crucified'; he ended up proving that crucifixion can be more or less continuous.

Even now, Greig's choice intrigues. For of all the World Series cricketers, none developed such simpatico with Packer. There was less than nine years' difference in their ages, and they seem to have caught each other's imaginations: Greig's growing interest in wealth was pricked by the money-man Packer, Packer's abiding fascination with winning was tweaked by the athletic adonis Greig.

Both, perhaps not coincidentally, had striven to prove themselves to despotic fathers: Greig's was 'always hypercritical of … every move I made'; Packer could not recall, where Sir Frank was concerned, 'a single occasion when he has praised me to my face'. There are aspects of both homage and rebellion to their actions: of Packer's in designing WSC, of Greig's in joining it. And it may have been the example of his boss's insouciance that has helped Greig cultivate his own.

The Age December 2003

John Wright

Trying

CRICKET IS A CHALLENGING GAME TO PLAY. It can also be, for coaches, an equally difficult game to watch. The best-laid plans can be squandered in a session; months of preparation might be rendered futile by injury or illness. Yet while actively involved in nothing, one is somehow accountable for everything.

India's coach John Wright bears this with a philosophical air. His appointment three years ago to replace Kapil Dev antagonised virtually every ex-Test player there: they rather fancied the role of hitting high catches and attracting a little reflected limelight themselves. His success is more noteworthy for his own playing career: where Indian teams have traditionally abounded in untapped ability, Wright was for fifteen years part of a New Zealand team that stretched a little talent a long way.

Wright the cricketer always reminded me of Roger Daltrey to look at, but he batted more like a John Entwistle, maintaining a steady tempo, minimising the histrionics. Sound in defence, neat off his legs and sometimes downright attractive square of the wicket, he played eighty-two Tests and was pushing thirty-nine when he retired.

His Test average of 38 is probably an accurate index of his skills. Unlike his outstanding contemporary Martin Crowe who made batting look second nature, Wright always radiated the aura of a terrific trier. Throughout his career, in fact, Wright admitted towards its end, he suffered acutely from nerves: 'I've tried to analyse why I get nervous because I've always thought it would be great to play clinically, like a surgeon, cutting your emotions off completely. But I'm not made that way temperamentally.' He had all the strokes, yet seldom licensed himself to execute them: 'I've often gone out saying

140

to myself that I'm going to smack it but not had the guts or conviction to see it through.'

No batsman in my memory, moreover, ever looked so thwarted on
getting out. Not that he was ever surly or truculent; instead, he would
go through a dumb show of disappointment, with himself and with
the weird, wonderful, confoundedly frustrating game he played. He
would shake his head and shut his eyes in grief, his mouth forming
inaudible curses as he mentally replayed his errant stroke.

What's interesting about Wright, however, was a flair, almost in
spite of himself, for rising above the crises of the day; he was, in all
other respects, an easygoing character who derived endless delight
from his career. Wright's 1990 autobiography, *Christmas in Rarotonga*,
is an exception in a generally naff genre: personal, warm, unselfconscious, and the only cricket book I know of dedicated 'to my thigh
pad'. Its introduction concludes with this disarming preamble:
'Assuming that anybody who has read this far has actually bought the
book, I can confess that what it does contain are the reflections of a
bloke who's played a fair bit of cricket with and against the best of his
time, in various parts of the world and, as he approaches the end of
his career and faces the appalling prospect of having to do some real
work, is trying to squeeze a few more bob out of it. Once a pro, always
a pro.'

Christmas in Rarotonga makes little effort to sell itself. The first
third is devoted, almost archaically, to Wright's county career with
deeply unfashionable Derbyshire, alongside comrades like David
Steele, an erratic runner who would farewell partners he ran out with
a sympathetic 'that's show business', and the Danish paceman Ole
Mortensen, who would respond to setbacks with a scream of 'Satan!'
There's relatively little in Wright's book about himself, and nothing
self-regarding: few cricket autobiographies can have contained quite
so many photographs of the author being bowled. There is instead a
palpable pleasure in characters met along the way, high and low. Star
players are studied with the same sympathetic eye as bat-makers,
pavilion attendants, scorers and umpires. The flatulence of Aussie
ump Cec Pepper, we learn, always enlivened life for non-strikers. 'Just

kick that one to the boundary, will you John?' he once asked Wright after a particularly sonorous blast.

In recalling his Kiwi contemporaries in those ragged but rugged teams of the 1980s, again it is affinity for the undersung that shines though. He relates how Chris Kuggeleijn coped with a pair in India by removing his pads, lighting a cigarette and announcing: 'I think I'll go up to the scorer's box and check out my run charts.' He describes the irascible John Bracewell putting an Adelaide cabbie in his place. 'Didn't you used to dig graves?' asked the driver. 'Don't you still drive a cab?' snapped Bracewell. Perhaps it is this sense that cricket must be enjoyed in order to be endured that equips Wright in his current role. He brings to a team that wants for nothing a capacity for pleasure in everything. I wonder, in fact, how Wright the coach would have handled Wright the batsman.

The Age December 2003

Chris Tavaré
Gentle Man

SOME YEARS AGO, I adjourned with a friend to a nearby schoolyard net for a recreational hit. On the way, we exchanged philosophies of cricket, and a few personal partialities. What, my friend asked, did I consider my favourite shot? 'Easy,' I replied ingenuously. 'Back foot defensive stroke.'

My friend did a double take and demanded a serious response. When I informed him he'd had one, he scoffed: 'You'll be telling me that Chris Tavaré's your favourite player next.' My guilty hesitation gave me away. 'You poms!' he protested. 'You all stick together!'

Twenty years since his only tour here, mention of Tavaré still occasions winces and groans. Despite its continental lilt, his name translates into Australian as a very British brand of obduracy, that Trevor Baileyesque quality of making every ditch a last one. He's an unconventional adoption as a favourite cricketer, I'll admit – yet all the more reason to make him a personal choice.

Tavaré played thirty Tests for England between 1980 and 1984, adding a final cap five years later. He filled for much of that period the role of opening batsman, even though the bulk of his first-class career was spent at numbers three and four. He was, in that sense, a typical selection in a period of chronic English indecision and improvisation, filling a hole rather than commanding a place. But he tried – how he tried. Ranji once spoke of players who 'went grey in the service of the game'; Tavaré, slim, round-shouldered, with a faint moustache, looked careworn and world-weary from the moment he graduated to international cricket.

In his second Test, he existed almost five hours for 42; in his third, his 69 and 78 spanned twelve hours. At the other end for not quite an hour and a half of the last was Ian Botham, who ransacked 118 while Tavaré pickpocketed 28. As an ersatz opening batsman, he did not so much score runs as smuggle them out by stealth. In the Madras Test at the start of 1982, he eked out 35 in nearly a day; in the Perth Test at the end of 1982, he endured almost eight hours for 89. At one stage of the latter innings, he did not score for more than an hour. Watching on my television in the east of Australia, I was simultaneously aching for his next run and spellbound by Tavaré's trance-like absorption in his task. First came his pad, gingerly, hesitantly; then came the bat, laid alongside it, almost as furtively; with the completion of each prod would commence a circular perambulation to leg to marshal his thoughts and his strength for the next challenge.

That tour, I learned later, had been a peculiarly tough one for Tavaré. An uxorious man, he had brought to Australia his wife Vanessa, despite her phobia about flying. Captain Bob Willis, his captain, wrote in his diary: 'He clearly lives every moment with her on a plane and comes off the flight exhausted. Add to that the fact that he finds Test cricket a great mental strain and his state of mind can be readily imagined.' You didn't have to imagine it; you could watch him bat it out of his system.

Tavaré could probably have done with a psychiatrist that summer; so could I. Our parallels were obvious in a cricket sense: I was a dour opening batsman, willing enough, but who also thought longingly of the freedoms available down the list. But I – born in England, growing up in Australia, and destined to not feel quite at home in either place – also felt a curious personal kinship. I saw us both as aliens – maligned, misunderstood – doing our best in a harsh and sometimes hostile environment. The disdain my peers expressed for 'the boring Pommie' only toughened my allegiance; it hardened to unbreakability after his 89 in Melbourne.

Batting, for once, in his accustomed slot at number three, Tavaré took his usual session to get settled, then after lunch opened out boldly. He manhandled Bruce Yardley, who'd hitherto bowled his off breaks with impunity. He coolly asserted himself against the pace bowlers who'd elsewhere given him such hurry. I've often hoped for cricketers, though never with such intensity as that day, and never afterwards have felt so validated. Even his failure to reach a hundred was somehow right: life, I was learning, never quite delivered all the goods. But occasionally – just occasionally – it offered something to keep you interested.

Wisden Asia October 2003

Len Pascoe
Fast Times

FAST BOWLING, SAY MODERN SPORTS SCIENTISTS, is all about fast-twitch fibres: muscles whose contraction generates force in an instant, and which the body recruits for activity requiring explosive energy. Their fast-twitch endowment explains well how the likes of Malcolm Marshall or Michael Holding, slight and sinuous, achieved their break-neck velocities. In other cases, however, it seems altogether redundant.

Len Pascoe looked to have as many fast-twitch fibres as your average tree trunk. Yet you never wondered where his speed came from: it was all brute force and ignorance. Looking at film of him now, you are reminded of how fast-bowling bodies have changed since they started taking out gym memberships, scorning beer in favour of electrolyte drinks and playing in their own rock bands. Nobody would have mistaken Pascoe for a high-performance athlete – or, indeed, a pop star. Standing 187 centimetres tall, with a fighting weight of about ninety kilograms, he was constructed on the lines of a circus strongman: thick neck, broad shoulders, barrel chest, big backside.

His favourite cricketer, Pascoe once said, was the Englishman Fred Trueman, and he assuredly belonged to the Trueman tradition, built to bowl all day and carouse all night. Like a Trueman, too, he had the histrionics to go with his hostility. He pawed the ground, flared his nostrils, tore in as though the devil was at his heels, and banged the ball in short – powerfully, persistently, and sometimes it must be said pointlessly, short.

The line he was allotted in 'C'mon Aussie C'mon' made him sound like a golfer – 'Pascoe's making divots in the green', so as to rhyme with 'Lillee's pounding 'em down like a machine' – but you only had to watch him bowl an over to know exactly what the Mojo librettist was on about. Pascoe was not only fast, he looked fast. He strained for that extra knot. He flogged himself for that extra over. While his chest-on action was simple and economical, he uncoiled like a clockwork toy wound as tightly as it would go. Then, often as not, came the trademark Pascoe glare, from deep-set, dark-brown eyes beneath a heavy brow. A characteristic pose was standing half-turned at the end of his follow-through looking severely back over his shoulder; he had a nose that would have suited carving into Mount Rushmore.

Pascoe personified so many fast-bowling clichés, in fact, that it was almost like a parody of the art, except that a palpable character lurked behind it. His aggression was never playing to the gallery. In one of the malapropisms for which he developed a dressing-room reputation, he once explained that 'a leopard can't change his stripes'. He would have bowled the way he did on his own with a tennis ball in a darkened room against a wall – and probably told the wall it was nothing but a bloody fence anyway.

Pascoe's volatility and volubility were sometimes explained by reference to his Yugoslavian parentage. He first represented the New South Wales Colts as nineteen-year-old Len Durtanovich, fresh from Punchbowl High School; English opponents privately nicknamed him 'the Separatist', seeing in him a species of Balkan fanaticism.

There was, though, just as much suburban Sydney about Pascoe. He was a council arts and crafts organiser. He drove a V8 Torana. He liked steak and chips, westerns and fishing. As a young man, he skipped a whole summer to go surfing on the north coast of New South Wales. In some respects, in fact, Pascoe was less reminiscent of a Test fast bowler than the club cricket tearaway who bounces, eyeballs and sledges you all Saturday afternoon. When he and his Punchbowl pal Jeff Thomson appeared for Australia, you might have been watching them play for Bankstown. Representing their country was pretty cool; but at least as much fun was bowling quick, walking tall and giving those batting pissants a bit of hurry up.

As attestation of one's own memory, statistics are often a bit disappointing. Batsmen you revere turn out to have had a rather leaner time than you recall; bowlers you remember as irresistible prove to have gone around the park a few times. Pascoe belongs to that far smaller subset of cricketers whose figures surprise by their excellence. In fourteen Tests, Pascoe struck 64 times at 26.06 runs per wicket, 5/59 in 1980's Centenary Test at Lord's being his best day out. His 53 scalps in twenty-nine one-day internationals cost only 20.11 runs each, and took less than five overs each to obtain.

He kept, what's more, some fair company. Australia abounded in pace-bowling talent in the late 1970s and early 1980s: it was the age not merely of Lillee and Thomson, but of Hogg, Lawson, Alderman, Rackemann, Dymock, Prior, Malone. And if the flesh sometimes made you wonder, the spirit was never other than willing.

The Age February 2004

Arjuna Ranatunga
Stumbling Block

AUSTRALIAN CRICKETERS RARELY GIVE MUCH AWAY in their public comments, preferring to walk softly and carry a big bat. Where Sri Lankan captain Arjuna Ranatunga is concerned, however, they'll make an exception.

'An abrasive customer who deliberately gets up the opposition's nose,' contends Mark Taylor. 'Without doubt the most difficult cricketer I've met or played against,' gripes Shane Warne. 'The main stumbling block in [Australia–Sri Lanka] relations,' concurs Ian Healy. Gee fellas, get off the fence and tell us what you really think.

Of course, it's an incomplete assessment. In his eighteenth year at the top, after almost 13,000 runs in eighty-two Tests and 250 one-day internationals, Ranatunga is international cricket's most enduring serviceman. Surviving so long anywhere would be a feat. Surviving in a country where the cricket politics carries on with Medici-like intrigue testifies to some astounding skills of perseverance and self-preservation.

Make no mistake, Ranatunga can bat. Former Sri Lankan coach Dav Whatmore enumerates his gifts: 'Good eye, solid defence, plays late, strong wrists, very strong physically, a bloody good player.' And Ranatunga can captain. 'A leader of men,' comments Ian Chappell. 'He doesn't take any rubbish from anyone, not the opposition, nor his own administrators.' So what is it about this unprepossessing 173-centimetre figure with his cummerbund of tubbiness that checks Warne's stride like chewy on the bottom of his Nikes?

Put it this way. Ranatunga's parents wanted him to be a doctor. So he is, kinda: up your nose, in your ear, under your skin simultaneously. He goads bowlers with cheeky strokes, goads fielders into overthrows with strolled singles, goads captains with continual requests to the dressing room, goads umpires with spontaneous injuries and pleas for a runner. 'Often when he is allowed a runner,' Warne complains, 'Ranatunga will smirk at us to rub it in.' Backhanding the Australians for sledging, ordering his team not to shake hands after a bitter one-day final in Sydney, dismissing Warne as media hype and the Waugh twins as overrated: the Serendip Lip is cricket's number one *agent provocateur.*

Apprehensive about a repeat of their team's unruly tour three years ago, the Board of Control for Cricket in Sri Lanka is investing in spin: a spin-doctor, to be precise, in the form of a Melbourne PR agent to intermediate with local media. But his workload will still hinge on someone who, to Taylor's cohorts, makes Saddam Hussein look like Nasser Hussain.

*

Given the BCCSL's apparent embrace of the power of positive PR, I feel confident of an audience with Ranatunga. When the BCCSL's CEO foresees no problem and asks only a formal written request, I feel even more confident. The BCCSL's CEO, after all, should know Arjuna's mind: he is his elder brother and former Test opener Dammika. In fact, this is one mighty clan: two younger brothers (Sanjeeva and Nishantha) have also represented Sri Lanka, another (Prasanna) is on the BCCSL executive committee, while Dad (Reggie) is a Sri Lankan Freedom Party stalwart and now a junior minister. Sri Lankan cricket pundit Mahinda Wijesinghe likens the Ranatungas in stature and influence to the Graces, who towered over English cricket a century ago, with Arjuna a dusky, whiskerless W.G.

It wasn't always so. Cricket came to Sri Lanka as part of the colonial box-and-dice during the island's 132 years as part of the British Empire, and emerged with a pronounced English intonation. In august Colombo colleges like Royal and St Thomas, generations of boys like future Test captains Duleep Mendis and Ranjan Madugalle were trained in cricket's technical precepts and its sustaining moral code.

Ranatunga is representative of a different elite: upwardly mobile middle-class Sinhala speakers who began making good in politics, business and sport after independence in February 1948, often after attending one of Colombo's two premier Buddhist colleges, Nalanda and Ananda. Sri Lanka's first Test captain Bandula Warnapura was an alumnus of the former, as have been Asanka Gurusinha, Roshan Mahanama and Kumara Dharmasena; the brothers Wettimuny and Ranatunga are products of the latter.

An opposition has always existed in Sri Lankan cricket between old world and new. Visiting the island to coach in 1981, Sir Garfield Sobers commented on the game's peculiar hybrid feel: Sri Lankans had a 'very traditional approach' countervailed by 'something of the Caribbean spirit coming from their own nature'.

Anglophilial fair play was Sri Lanka's hallmark when it successfully petitioned for Test status in July 1981. Quizzed about sledging when his country toured England that season, Sri Lanka's assistant manager responded ingenuously: 'Sludging? What is sludging?' Arjuna and his comrades, however, caught up quickly: Sri Lanka won only two of its first forty Tests, and it was the manner as much as the actuality of its defeats that smarted. Such was the blizzard of abuse when an appeal for a bat-pad catch against Ranatunga was disallowed in Pakistan's inaugural Test in Sri Lanka at Kandy in March 1986 that umpires suspended play for half an hour. When he and Gurusinha scored freely in a Delhi one-dayer in January 1987, India slowed them by deliberately bowling only 44 of the mandatory 50 overs in the allotted time. Where quick bowling was concerned, the difference in attitudes was particularly acute. Richard Hadlee peppered the Sri Lankan tail with bouncers at Kandy, but Rumesh Ratnayeke fainted at the sight of blood when he broke John Wright's nose at Basin Reserve and had to be revived with smelling salts.

Making it still harder was the island's constant civil strife. Riven by conflict between the Sinhalese and minority Tamils in the northeast, Sri Lanka has spent more than half its history under emergency rule. Ranatunga has first-hand experience of the volatility of local politics: the family home was torched by political rivals in the early 1970s, leaving them with barely a stitch.

Conflict's impact on cricket – which has since the election of Julius Jayawardene in October 1982 claimed 50,000 lives – was doubly debilitating. On one hand, it checked Sri Lanka's integration with the cricket world: the country could host only a dozen Tests in its first decade. On the other, it heightened expectations of the national cricket team as a rare symbol of unity, and deepened political interference in the BCCSL. It has been run by a succession of senior cabinet ministers since its inception, and its selectors and selections must still meet Sport Ministry approval. Success brought extravagance: Jayawardene declared a national holiday in April 1986 when Sri Lanka won its first one-day trophy. Disgrace brought recriminations: Sri Lanka had six captains in its first six years of international cricket.

*

I'm looking forward to asking Ranatunga about this challenge of creating an authentic voice for Sri Lankan cricket. For this surely is his accomplishment: when his country carried off the World Cup in 1996, it was with a snap and a dash that owed nothing to the country's colonial inheritance.

I can't get through to Dammika Ranatunga for ten days, however, and he's confused when I do. 'You haven't received anything?' he says. 'I thought that had been done.' Still, he's helpful: if I fax him again, he'll do something that day. I do. He doesn't. But finally, late the following day, a BCCSL fax advises 'no objection' to my interviewing Ranatunga 'provided that he is agreeable'. It seems he is when he answers a Sharjah hotel room phone next day and, wary but polite, suggests a time a few days hence. I wish him luck for his next match in the Champions Trophy. He thanks me. We say goodbye.

Off-field, I'm told, such politeness is characteristic. A devout Buddhist, Ranatunga reveres his father and calls his coaches 'aiya' (elder brother) as a mark of respect. On-field, the determination emerges. Asanka Gurusinha recalls: 'To me, he was the best of partners to bat with. We would say: "There are only two of us against eleven others, so we must support each other." We would always talk to each other between overs, even during overs. If I played a bad shot, he would come down the pitch and say: "Come on Guru. Don't do that. We've gotta fight it out."' Ranjan Madugalle remembers batting with the twenty-one-year-old Ranatunga against Pakistan at Sialkot, and copping a fierce ball and mouth interrogation from Imran Khan. 'You concentrate on your batting, Ranjan,' confided Ranatunga. 'I'll handle the rest.' He did. Indeed, he needled Imran so ceaselessly that the Pathan prince promised him cricket-ball cosmetic surgery. 'That's okay,' Ranatunga retorted. 'You have the ball, I have a bat. Let's see who wins.'

When Ranatunga succeeded Madugalle as captain in October 1988, the plan was clear. 'We had a reputation for being cricketers who enjoyed the game and never grumbled,' says Ravi Ratnayeke, Ranatunga's first vice-captain. 'But we realised that, if we wanted to progress, we needed to win, to beat people at their own game. Arjuna was the right person. He brought a bit of muscle, a bit more cunt into the side ... Previously our captains had always come to games talking about how not to lose. Arjuna would say: "What's the point playing if you don't want to win?"'

*

I call at the agreed interview time. Ranatunga is asleep. His wife asks if I can call back in an hour. I do. Ranatunga isn't there, and isn't for the rest of the day. I leave messages. They aren't returned.

I call next morning. Ranatunga is asleep again. His wife suggests another time later that day. I agree but, when I call, there is again no answer. So much for PR.

I'm not altogether surprised. Where the media is concerned, Ranatunga has never been a Dale Carnegie disciple. His wilful streak and flair for brinkmanship have sometimes cost him in other areas too. His second Test tour as captain to New Zealand in January 1991 was soured by a stand-off arising when Dammika had his right thumb broken in Sri Lanka's first match. Given that the injury was unlikely to heal in the tour's five weeks, manager Stanley Jayasinghe

proposed that Dammika be sent home. Ranatunga stubbornly sought
to overrule him.

The dispute strained friendships. Jayasinghe, a Nalanda old boy
who'd played county cricket for Leicestershire, was well regarded.
Gurusinha recalls: 'To me, Stanley was a fantastic manager. And that
is where we fell out, Arjuna and I, because I supported Stanley. I don't
know whether Arjuna was doing it because Dammika was his brother.
But I thought that the decision was a matter for the manager not the
captain.' Sri Lanka excelled on tour, Aravinda de Silva and Gurusinha
both averaging more than 70, but subsequent investigation by a
retired judge of Ranatunga's 'haughty' attitudes and 'double standards' saw him reduced to the ranks. De Silva and Gurusinha led their
country for the next eighteen months.

The experience was not mellowing. Ranatunga fared well at first
when a changing of the BCCSL guard permitted his rehabilitation, but
again became mired in controversy during the Hero Cup in November
1993. Sri Lanka lost four of five games heavily and their vital anchor-
man Gurusinha to a torn adductor muscle. Invalided home, bizarrely
enough, Gurusinha found on his doorstep a BCCSL letter demanding
repayment of the balance of his tour fee: about $1400. Unable to pay
at once, he offered to refund the money in instalments, but was
promptly suspended.

With Gurusinha out of favour and de Silva out of form, Sri Lanka
was thrice routed by an innings in the subsequent Test series in India.
Then, even more bizarrely, Ranatunga and ten colleagues paused at
Colombo airport just long enough to fulfil their tour contracts before
returning to India for a series of benefit matches. Another BCCSL
inquiry pilloried Ranatunga and de Silva. Both were selected for the
subsequent Australasia Cup in Sharjah but, just before departure, de
Silva was mysteriously declared 'unfit' by the Ministry of Sports.
Ranatunga led a senior-player boycott which precipitated angry edi-
torials, hunger strikes and death threats for Ranatunga's replacement
skipper Roshan Mahanama and Gurusinha (for whom a friendly
businessman had now repaid his disputed tour fee).

This time, Ranatunga prevailed. Apparently by presidential decree, the BCCSL issued its captain a cringingly fulsome pardon recognising his 'outstanding contribution to Sri Lankan cricket'. The philosophy seemed to be that expressed by US president Lyndon Johnson when he included political nemesis Robert Kennedy in his cabinet: 'I'd rather have him inside the tent pissing out than outside pissing in.' The BCCSL regime then fell when the ruling United National Party were routed at August 1994 polls, and the election of Ranatunga's father and lavish editorial praise in the state-owned *Daily News* further buttressed his position. So who needs friends when so many other ways exist to influence people?

*

I contact Dammika Ranatunga again in Colombo a week after the debacle in Sharjah. Arjuna is now too busy for an interview. Can I fax a list of questions? I do. In the meantime, I visit Sri Lankan-born, Australian-bred Dav Whatmore who arrived in Colombo in May 1995 as national coach and brought a wealth of fresh ideas about physical and mental preparation for the business of international cricket.

He speaks warmly of Ranatunga. Though obviously no athlete, the Sri Lankan skipper didn't shirk Whatmore's exacting schedule of fielding, running and weight training. 'He tried hard,' says Whatmore. 'He wasn't winning any races, but he was always on time and certainly did all the work. In fact, he lost a lot of weight in the first few months I was there.' The discipline paid. Coming from behind in both cases, Sri Lanka won Test and one-day series in Pakistan, then the Singer Champions Trophy in Sharjah.

The subsequent tour down under was calamitous, with cocked-up ball-tampering allegations in Perth, cocked-arm mortification of Muttiah Muralitharan in Melbourne, and crocks jamming the door of Australian physiotherapist Alex Kountouri, including de Silva, Gurusinha, Mahanama, Manjula Munasinghe and the skipper himself (who broke a hand just before Christmas at the SCG). But both Ranatunga and Whatmore left feeling surprisingly confident about Sri Lanka's prospects in the subsequent World Cup on the sub-continent: on familiar surfaces and before different umpires, their XI aggregating more than 1000 one-day caps would give anyone a run for its money. Even Australia. But not, perhaps, particularly Australia.

One shouldn't exaggerate. Australians feel strongly about Ranatunga, but probably more strongly than Ranatunga feels about Australians. Australians were aggrieved three years ago that he wouldn't socialise with them after play. But Ranatunga has never socialised with opponents: he seldom ventures outside hotels on tour except to play. And there *were* occasions when Ranatunga showed a different side of himself: like speaking admiringly of Allan Border in Brisbane and visiting David Boon in Adelaide to wish him well on retirement and present him with a Sri Lankan tie. Indeed, Whatmore contends, Sri Lanka's captain has a sneaking regard for Taylor's team: 'I think Arjuna has a deep respect for Australia: they're a team that you tend to measure yourself against individually and collectively to see how you're going. It's just that, say, the way Shane Warne has been written up would act as an extra incentive for him. Arjuna would go out of his way to do well against him, to say to the other batsmen: "Hey, everyone's raving on about Warne. But we're not scared of him. We'll bloody show them." Because you're an island, and because globally you're looked at as a minor country, you've gotta stand up and make those sort of comments.'

No doubt both teams had points to prove in that Lahore final of 17 March 1996. No doubt either that Ranatunga outfoxed his nemeses. His pre-match rhetoric played ostentatiously on Australia's forfeiture of its opening match of the tournament in Sri Lanka, after escalating civil strife: 'They have avoided us once. They can't avoid us now.' But when the Australians batted, like angry men bent on retribution, there was no pace for them to hit: Ranatunga plied them with 37 overs of spin, including 17 of rinky-dink slows from de Silva and Sanath Jayasuriya claiming four economical wickets. And when the Sri Lankans followed, keyed-up but clinical, they exploited evening dew robbing Warne of grip and purchase: Ranatunga pongoed him for four and six in the forty-first over to put the result beyond doubt.

The satisfaction of the World Cup for Ranatunga, however, was less in the settling of a few short-term scores with Australia than in the climax of fifteen years sloughing off his country's old cricket skin. Sri Lanka had taken the tournament with an admixture of guts and guile, brazen strokes and brainy bowling, that was of 'their own nature'. And especially of Ranatunga's.

*

I'd like to have discussed this with Ranatunga but, as you may have guessed, didn't get the chance. The BCCSL promised reply to my questions 'in due course', but deadline impended after another fortnight, and 100 phone calls to Colombo and Sharjah ultimately yielded squat. I wasn't bothered overmuch: nothing obliged Arjuna to talk to *Inside Sport* in the first place. But this was seasoned with another reflection: nothing obliges Arjuna to do very much at all these days.

In the thirty-two months since winning cricket's global gourd, Sri Lanka has experienced another makeover. Whatmore quit in November 1996, successor Bruce Yardley in July 1998, and victory from behind against New Zealand and a maiden English Test win has been counterweighted by meek capitulation in South Africa and a wretched Champions Trophy in Sharjah two months ago. Ranatunga omnipotent reigns, most believing that Sri Lanka's captaincy is his as long as he wishes. But therein might lie the greatest risk to his abiding renown: not the likelihood of another skirmish in Australia, but the unlikelihood of one at home. Ranatunga is fit, having shed fourteen kilograms in the last year, but the game doesn't get easier for one in their thirty-sixth year, and it doesn't take much imagining to foresee another cyclical dip in Sri Lankan cricket as stalwarts like de Silva (33), Mahanama (32) and Hashan Tillekeratne (31) pass their peaks. Ranatunga has established Sri Lankan cricket's foundations, but will he let others build on them? Will he be able to bequeath the work of half his lifetime to another generation? Or will Sri Lanka reprise that old joke about Soviet Russia?

First lackey: The dictator is dead.

Second lackey: Who's going to tell him?

Inside Sport January 1999

Lawrence Rowe
The Enemies of Promise

A BATSMAN AVERAGING 43 FROM thirty Test matches would justly be considered an accomplished practitioner of his art. Yet what if, after a dozen Tests, he had been averaging 73? And what if, after one, he had been averaging 314?

Lawrence Rowe's respectable but nondescript Test record obscures a unique career trajectory: no more extreme example of disappointed hopes can be found in international cricket. Some in the Caribbean still believe it has produced no-one with such natural gifts; none can describe him without a wistful shake of the head.

Rowe's outsized average after his Test debut arose from what remains a record performance, against New Zealand in February 1972: 214 and 100 not out at Sabina Park, following 227 against them on the same ground for his native Jamaica. Nor was this merely a statistical freak. Rowe was one of those unusual batsmen to whom a completely orthodox technique came quite naturally, as though cricket was always in him, waiting to be revealed. His bat was dismayingly straight, and his signature shot was off the back foot through the covers – always indicative of time to spare. Critics likened him to Frank Worrell, and it was Worrell's father-in-law who volunteered the loveliest line of all; he commented that God seemed to have put a bat in Rowe's hands and counselled him: 'Go thou and bat.'

Rowe could belabour attacks violently. Early in his innings against England at Kensington Park in March 1974, he received a bouncer from Bob Willis. He smashed it flat into the stand at square leg; it travelled most of the way at little more than head height. His 302 took only 430 hectic deliveries, with thirty-six further boundaries. More often, Rowe exuded ease. He was said to hit the ball just hard enough to reach the boundary, so that fielders always had to chase but never quite caught anything. And, completing the image, he whistled a tune after each attacking shot, as though to complement the music of his bat.

With Viv Richards still to come, Rowe was in his time the West Indian glass of fashion. 'He became our hero,' said Desmond Haynes. 'He had such … such style.' Michael Holding still believes Rowe 'the best batsman I ever saw', and that he 'could not imagine anyone ever batting better or being able to'. Even then, however, Rowe's career had reached its zenith, and his regression to the mean would be a mix of hauntingly bad luck and a superstitious and suggestible nature.

When he was recruited to play county cricket by Derbyshire, he was immediately routed by hay fever and headaches: of all things for a cricketer to be allergic to, he suffered a reaction to grass. He made runs, but uncomfortably and miserably.

When he then joined the West Indies on tour in India in October 1974, Rowe seemed to fall apart. He failed, tamely, thrice; even in the nets, he was stilted and fallible. Sent home for treatment of a stye on his eyelid, he was discovered by an ophthalmic surgeon to have eyesight better than 20/20, and could literally read the maker's name on the optical chart. But he was also suffering pterygium, a disease involving vision-blurring growths: they had almost completely covered his right eye and were on the way to obscuring vision in the left.

A remedial operation damaged his eyesight; contacts were prescribed but, in those days before soft lenses became standard, caused his eyes to water profusely. And for a batsman whose game had been built on an eagle eye, the deterioration in Rowe's sight must have been traumatic; West Indian cricket's most exhaustive chronicler, Michael Manley, described him as 'transfixed by misfortune'. But Rowe's afflictions also caused him to look more deeply into himself – and there he found hitherto-unacknowledged weaknesses.

Holding's autobiography contains a fascinating vignette of the young Rowe, describing a country social game where he was the special guest. Folk came from near and far to watch but, amid acute disappointment, the great man refused to bat: 'There had been rain, the pitch was damp and he protested that, in the conditions, he could not be the Lawrence Rowe the people were expecting.'

In the second half of the 1970s, the challenge was still more exacting: he could not be the Lawrence Rowe that Lawrence Rowe expected. He still touched batting heights reserved for very few: his virtuoso 175 for the West Indians in the VFL Park Supertest in January 1979 was arguably the best batting act seen under the Packer big top. Yet setbacks seemed to resonate with all Rowe's doubts. 'When things started to go against him, he would blame his failures on everything but himself,' Holding wrote. 'It reached the stage where he became so paranoid … that he was reluctant to go out to bat for fear of failure.' Once, returning to the dressing room after snicking a steepling lifter from Jeff Thomson, Rowe was heard to protest: 'Not even God could play that!'

Deserted, Rowe chose to desert. When agents of South African cricket came calling in December 1982, he volunteered to lead the rebel West Indian team there: a route to short-term prosperity and long-term pariahhood. It was an impulsive decision, built no doubt of greed, but also of disillusionment. If his career was to be one of disappointed expectations, Rowe perhaps felt, he might as well dash them utterly.

The Age March 2004

Present Discontents

World Series Cricket
Interesting Times

TWENTY-FIVE YEARS AGO, on 2 December 1977 to be precise, Australian cricket lovers turning on their television sets had for the first time a choice in their bill of fare. Live from the Gabba on the Australian Broadcasting Commission came the soothing sights and sounds of a traditional Test, the first of a series against India. Live from Melbourne's VFL Park on Channel Nine, meanwhile, came the unfamiliar images of what purported to be a revolutionary new variant on the game: a Supertest, brought to you by World Series Cricket.

The play itself, between an Australian team led by Ian Chappell and a West Indian outfit captained by Clive Lloyd, didn't actually look all that different. The ball was red. The players wore white, and as yet sported caps rather than helmets. The Australian headgear, though, was gold not green, and it was such distinctions of detail that mattered. There were no traditions here. The ground, usually the preserve of Australian Rules football, had been converted by the installation of a pitch grown in a greenhouse. The television coverage, rather than relying on the usual two cameras, used eight, with extensive reliance on video replays. Microphones embedded in the ground near the stumps captured the players' grunts and the wickets' rattle; a boundary interviewer even solicited their post-dismissal musings. Critics were already calling this a pirate enterprise: its symbol, a stylised set of black stumps partially enclosing an outsized red cricket ball, would become the game's equivalent of the skull and crossbones.

Cricket had been cleft in twain almost six months. The first plans for WSC, and the first international cricketers recruited by the agents of its impresario Kerry Packer, had been revealed in April 1977. The principles seemingly at stake – love of country versus love of money,

a century of tradition versus spontaneous spectacle – had been endlessly debated. But until that December morn, the rivalry's implications had been obscure. Packer's original objective, indeed, had not
been to introduce an alternative brand of cricket at all. His eyes were
on the prize of exclusive Test match broadcasting rights in Australia;
WSC was merely a roundabout way of bending the Australian Cricket
Board to his will. Now it was a twin-match, twin-tour, twin-channel
reality. 'The public will decide,' pronounced the editor of *Wisden*,
Norman Preston.

The public issued what looked like a decision that very day. Where
there were no traditions, there were also no spectators. While about
12,000 attended the Brisbane Test, fewer than 500 were scattered
round the concrete tiers of VFL Park where space could be found for
80,000. Packer had more stars than Broadway: the Chappells, Dennis
Lillee, Rod Marsh, Doug Walters, David Hookes versus Lloyd, Viv
Richards, Joel Garner, Michael Holding, Andy Roberts, with Tony
Greig, Barry Richards, Mike Procter, Imran Khan and Asif Iqbal in the
wings. But for what, punters pondered, were they playing? It clearly
wasn't for their country. It looked, uncomfortably, as though they
might be playing for money.

The story of WSC is strewn with useful lessons, and this one
counted, and has continued to count: that patrons seek some transcendent value in sport. Money alone won't do. They understand that
the labourer is worthy of his hire. They may even obtain a *frisson*
from the sheer vastness of a modern athlete's earnings. But they don't
support sport because they perceive it as a group of people earning a
living. And they don't believe that an athlete being paid a multiple
of his previous salary will be trying commensurately harder and
playing proportionately better. In this, they are actually more in tune
with sportsmen and women than most administrators and managers.
Nobody will undo sport's stealthy permeation by money, but introducing money to a sporting ecosystem cannot help but strain the
bond between spectator and spectacle: the wide open spaces on
Australia's National Rugby League terraces, attesting patrons' disillusionment with players who took the Murdoch shilling, are

reminiscent of nothing more strongly than those seen during WSC's first season.

Where Packer went right and Murdoch wrong isn't entirely about judgement. Packer had the huge advantage that sport twenty-five years ago was dirt cheap. As Lamar Hunt's father had warned him when they set up the World Championship Tennis breakaway in the early 1970s: 'If you're not careful Lamar, you'll go broke in a hundred years.' The Murdoch irruption on rugby league came when the game was already fully priced – his recent complaints about the inflation in the value of sporting properties seem as perverse as George Soros' denunciation of currency speculators.

Packer was smart enough, though, to realise that the money was useful to him only insofar as it obtained talent. The selling of WSC required more old-fashioned devices. The following season saw the launch of the most successful marketing campaign in cricket's history – the defining one, in fact. The strains of the 'C'mon Aussie C'mon' chant devised by the admen at Mojo can still set purists' teeth on edge, but none could dispute its efficacy in arousing the patriotic nerve of Australian cricket fans. Then, crucially, the WSC Australians began winning, including a night-time limited-overs match at the Sydney Cricket Ground on 24 November 1978 that drew more than 50,000.

Limited-overs cricket played at night with a white ball in front of dark sightscreens had been an innovation of WSC's first season, coming about for no reason other than that VFL Park had been equipped with light towers for night football. Now it became a WSC motif, especially after 17 January 1979 when a one-day match was staged for the first time in coloured raiment: the WSC Australians appeared in a burst of wattle gold, the WSC West Indies formed a reef of coral pink.

In the meantime, an underpowered Australian Test team was being steadily overwhelmed by Mike Brearley's visiting Englishmen, with a disastrous impact on the official exchequer. The establishment's defiant rhetoric had tended to conceal how poorly equipped it was to deal with a rival cricket promoter. It had only one source of sizeable revenue (Test cricket), from which to fund a set of cost centres (first-class, club and junior cricket); sponsorship and broadcasting monies were as yet paltry. That source of revenue, furthermore, had been predicated on a monopoly market position, which no longer pertained.

Packer, meanwhile, enjoyed the advantages of vertical integration. He was paying his cricketers big bucks – or, at least, bigger bucks than they'd been accustomed to. But he obtained for his bucks the additional bang of lots of popular, cheap, long-duration television for his summer schedules. And unlike the Australian Cricket Board, with its obligation to be all things to all fans, Packer had no need to stage matches outside the big markets of the eastern seaboard if he didn't feel like it; in its second season, WSC didn't visit Adelaide or Perth at all.

It couldn't go on. As Graham Yallop's callow XI succumbed in the Sixth Test at the SCG, a lone voice was heard calling from the Hill: 'Change the channel!' Which is what the Australian Cricket Board proceeded to do. From February through March 1979, WSC and the ACB worked towards a rapprochement under which Channel Nine obtained the broadcasting rights it had sought from the beginning. And they all lived happily ever after. Or so the story goes ...

World Series Cricket's historical impact has been in dispute for most of the last quarter century. Some have maligned it as the end for cricket as we knew it. Others have celebrated it as the beginning of cricket as we know it. For certain, cricket has never evolved so far so fast. Changes were wrought to the game's institutional structures in two years that might have taken ten or more. The one-day international was popularised, and the tri-cornered tournament made a feature of every subsequent Australian summer – a fashion that spread first through Asia, then to England.

Night cricket, coloured clothing and drop-in pitches were pioneered. Ditto helmets, which spread like mushrooms after rain when David Hookes had his jaw broken by Andy Roberts in December 1977. Administrators came to recognise television rights as an important revenue source. Cricketers became more cognisant of their market value. Broadcasters awoke to sport as popular mass entertainment, cheaper at that stage than just about everything except the evening weather. Television, too, ceased merely to be the game's silent witness. One of the features of Channel Nine's coverage was the narrative it imposed on the play, drawing attention to this and that, laying out the issues of the day even as it monitored the events. Watch how necks crane to study the big screen replay when something happens at a big cricket arena today – making up for the wavering attention on the play itself – and you realise how the style of coverage conceived by Packer's lieutenant David Hill has both enriched and impoverished our understanding of the game.

Over the last couple of years, in fact, it's become harder to raise more than two cheers for WSC. It represents the beginning of cricket's steady commodification. As Dr Greg Manning has written, Packer paid $12 million 'not to buy the cricket but to turn the cricket into something he could buy. The real meaning of his victory was that the game would never again be beyond price.'

Because professionalism was launched so rapidly, it has ramified in ways quite unforeseen twenty-five years ago. What happens, for example, if the highest bidder for your services should turn out to be the agent of an abhorrent regime? By promulgating the idea that a professional cricketer 'needs to make his living as much as any man' – as Justice Slade put it in his landmark High Court judgement in favour of the Packer organisation in October 1977 – WSC set the scene for a decade of steadily shabbier rebel tours to South Africa. What happens, too, if the highest bidder for your services should turn out to be a bookmaker? Then the only obstacle to your complicity is conscience – not a quality that has abounded in recent times. We live today, to use the famous Chinese curse, in interesting times. WSC helped make them so.

Wisden Cricket Monthly January 2003

The Qayyum Inquiry
The Incredible Exploding Cricket Team

MEET THE INCREDIBLE EXPLODING CRICKET TEAM: the batting, bowling and bookmaking Pakistan machine. Where players can be reprimanded for taking wickets and making catches, where bookies walk into hotel rooms with underpants full of banknotes, and where affidavits are signed as routinely as souvenir bats.

By now, you'll have heard of Justice Malik Muhammed Qayyum's report, recommending life bans on former captain Salim Malik and pace bowler Ata-ur-Rehman, and also fines for Wasim Akram, Mushtaq Ahmed, Inzamam-ul-Haq, Saeed Anwar, Waqar Younis and Akram Raza. But there's more. The compendious witness statements collected by the Lahore High Court provide a vivid picture of a team that's introduced many new techniques to cricket's textbook: notably the finger point, backside cover, sideways glance and face save. 'Is waqt sab ko sab par shaq ho raha hai,' mutters Saeed Anwar on one of the taped telephone conversations which wicket-keeper Rashid Latif presented to Qayyum: 'At this moment everyone is suspecting everybody else.' For when someone in Pakistan mentions suspect actions, they're not thinking of Law 24.

'We have to lose the match'

Pakistani cricket in disarray. Wasim Akram booted out as captain. Team-mates accusing one another of treachery. They could be headlines from today, but they were also the headlines in January 1994 when Pakistan's president Farooq Ahmad Khan Leghari dissolved his country's cricket board and appointed an ad hoc committee: chairman Javed Burki, Arif Ali Abbasi and Zafar Altaf. Abbasi testified before Qayyum that he persuaded his colleagues to crown Javed Miandad skipper for a forthcoming three-Test tour of New Zealand: 'However, next morning I was overruled and Salim Malik was made the captain.' So it was that one crisis begat another: if the inquiry's evidence is to be trusted, this overnight change-of-mind was the worst miscasting since Arnold Schwarzenegger made *Kindergarten Cop*.

The new reign dawned promisingly: Pakistan won the first two Tests easily. Then, however, they dropped the last, permitting New Zealand to overhaul 324 in the fourth innings. And the day before the last of five one-day matches in the BNZ Series – which Pakistan had already won – some peculiar things started to happen.

According to opener Aamir Sohail, team manager Majid Khan received information that the next day's game in Christchurch would be lost: 'Consequently he had banned all the telephonic calls to the players and informed them that henceforth all the telephonic calls would be routed to him.' But a 'telephonic call' got through to Rashid Latif all right. He testified that it was a summons to Malik's room where – in the presence of Waqar Younis, Inzamam-ul-Haq, Akram Raza and Basit Ali – the skipper offered him Rs 10 lakh ('lakh' means 100,000) to play badly. 'I told him that I would think over the matter,' swore Latif. 'In the next morning during the play when I took a catch of a batsman from New Zealand, Salim Malik came to me and reprimanded me and reiterated that we have to lose the match. During the water break I told Salim Malik that I was not party to the match-fixing as I decided during the night not to accept any money.'

In fact, Pakistan had already batted by this stage, scrounging 9/145. And Latif had another problem proving his story, thanks to television. Malik's 'reprimand' didn't go to air, the broadcaster having thrown to a commercial at the fall of the wicket. Another reason to put cricket back on the ABC ...

When plans were also threatened by Christchurch's inhospitable climate, Latif continued, Pakistan's bowlers had to donate wides; two sets of four flew past him to accelerate defeat. Akram didn't finish his spell – Latif reckons he feigned his shoulder injury – and

Ata-ur-Rehman bowled particularly badly. Or was there a reason for that, too?

Poor Ata. Almost two years ago he told the so-called Probe Committee, which led to Qayyum's appointment, that he had been contacted between 8 and 9 p.m. the night before the one-dayer by Wasim Akram, who offered him Rs 3–4 lakh for 'doing a favour' i.e. bowling a few timely pies. Ata obeyed, then had a pang of conscience: 'Thereafter my mother fell ill and my sister was operated upon and my conscience pricked me with the result that I stopped becoming a party to it.'

Since then, however, he hasn't known whether he's Ata or Mata. He withdrew the statement. Then he said his withdrawal was under duress. Then, under cross-examination, he said he'd been put up to it by Aamir Sohail. Ata is presently pro for the English league club Blackburn Northern, and you wonder whether at the bar after play he doesn't spend half an hour agonising over whether he's been coerced into his pint of bitter and cheese and onion crisps. For however badly he bowled in Christchurch, Ata finished up with an absolute jaffa: a reverse-swinging yorker into his own foot. He's been charged with perjury. But as the merely suspicious does not a case make, Qayyum had to conclude resignedly: 'There were misfields, there were wides. The batting collapsed. But then again that is the Pakistan team.'

Greenbacks and y-fronts

When rumours of a fix shadowed Pakistan's Australasia Cup final in Sharjah on 22 April 1994, coach Intikhab Alam took matters into his own hands, punting the $US10,000 losers' cheque on a win. He must have congratulated himself when the team coasted to a 39-run victory. Then, after a Test and one-day series in Sri Lanka, Pakistan had a couple of weeks off before the team entered the Singer Trophy tournament there.

Time for a bit of a break. Malik wanted to go home to Islamabad beforehand: to a wedding, Javed Burki reckons Malik told him; because 'my son had fallen and suffered an injury and I wanted to attend to him' testified Malik; to see a bookie says Latif.

Latif was also popping back to Pakistan and told the commission that, en route to Karachi, Malik lost his bags; no surprises there, as they were flying Aeroflot. The shock came when the missing article turned up, Latif recollecting: 'In his baggage, he was having 50,000 Sri Lankan rupees which he had won by gambling among the players.'

Latif testified further that, on returning to Sri Lanka, he saw two Lahore bookies: Saleem Pervaiz aka Pejii, who'd once opened for Pakistan, and one Aftab Butt. And when Pervaiz was called, he gave some gob-smacking testimony about Pakistan's ensuing match with Australia: 'I went to Sri Lanka on the asking of Mushtaq Ahmed who said I should come to Sri Lanka so as to see some matches, and maybe something comes of it.' Given that this 'something' might be a bit pricey, he came prepared: 'I carried [US]$100,000 with me. I had taken this money to Sri Lanka as I knew that the team is going to sell the matches to Carry Packer [sic] or to Bakhatar [sic] and therefore I thought why I should not try.'

Say what? Presumably Pervaiz was referring to Abdul Rehman Bukhatir, who runs cricket in Sharjah, and Kerry Packer, who runs about half of the world. This being so, let us stress that their links to match-fixing are unknown.

Now, Mr Pervaiz, please continue: 'I handed over the dollars to both to them [Malik and Mushtaq] who were together in their room at the hotel. The two players had contacted me directly in this connection. They had asked for a larger amount but I told them that I have only 1 lakh dollars.'

Bingo. But there was some trouble with this jaw-dropping evidence: the witness. In Sydney, they'd call Pervaiz a 'colourful racing identity'. He confessed to a couple of prison stretches, and to a murder charge (which he'd gotten off). And details of his story kept changing, particularly about the presence or otherwise of the aforementioned Aftab Butt: a 'businessman' who, in one version of Pervaiz's story, carried the 100k in his underpants. The old greenbacks in the y-fronts trick, eh? Butt himself, meanwhile, remained an if. Qayyum went looking, presumably for someone with unusually capacious jocks, but he is, at time of writing, still at large. As it were.

Singer sew-up

Cricket's a 'funny game', but they don't come much funnier than Australia versus Pakistan at the Sinhalese Sports Club on 7 September 1994. Australia extracted 7/179 from their 50 overs, thanks largely to a thirty-eight-minute 46-run seventh-wicket stand between Ian Healy and Shane Warne. As Pakistan's reply commenced, Sohail claimed: 'When we were going out to bat, Saeed Anwar said that I have heard something that a match is fixed therefore we should bat carefully. I inquired the reason. He said we are going to lose the match.'

Pakistan were cruising at 2/80 when two other curious exchanges occurred. Akram's mobile phone rang and he was heard to confide that he 'did not know' to the caller: to Latif it sounded like the inquiry concerned a fix. Then twelfth man Zahid Fazal carried a message to Saeed, at the time a sparkling 43 not out: Javed Burki says Saeed later confided to him that it was an instruction to self-destruct. Whatever the case, Saeed retired with 'cramps', which then appeared to be contagious: Malik made 22 off 51 balls, Inzamam-ul-Haq 29 off 69, Akram 16 off 33, Basit a 13-ball duck. What had begun as a run chase became a run plod and fell 28 short.

Burki asked Saeed for a written statement about the Fazal message: 'At that time I was staying at the Pearl Continental. Saeed Anwar promised to come over and do the needful. Later on Saeed Anwar informed me that he could not do the needful as promised because his brother was threatened with dire consequences if Saeed Anwar came over to me and delivered the statement in writing.' Saeed confirmed before Qayyum that the frighteners had been put on his brother, but denied having been told to get out; the message had been to play with 'care and caution'. You wonder why, of course, such a vastly experienced batsman had to be told anything.

That wasn't all. Late that night, Intikhab recalled, he received a 'furious' call from someone claiming to have lost Rs 40 lakh on the game and that four or five players had been 'bought'. When he summoned Malik, Waqar and Basit, the last 'confessed before me that he was involved in match-fixing' (something Basit later denied before Qayyum).

According to Intikhab, Asif Iqbal, the former Pakistan captain now at Bukhatir's right hand in Sharjah, then flew in from Washington 'though he had nothing to do'. He imparted to Intikhab a surprising confidence: 'He told me bluntly that bookies had lost Rs 40 lakh and they wanted to recover same at any cost. I had known Asif Iqbal since very long and was shocked to hear what he said to me.' At the probe committee, Intikhab testified further: 'Asif Iqbal ... took some of the Pakistan players and started talking to them in the whispers.' Asif Iqbal has since been named as a match-fixing go-between; a charge he has denied in *Gulf Today* in anything but 'the whispers'.

Then there was Mushie. The little leggie actually had a useful game, dismissing both Waughs for 34 in 10 overs; nice in any circumstance, and particularly handy in this case. But when he appeared before Qayyum, he made what the judge called an 'interesting' remark. As counsel steered round to the Singer Cup, he suddenly protested: 'But I was OK in that match!' As the actual game hadn't been cited, he seemed to have bowled himself a bit of a googly; Qayyum recommended that Mushtaq be fined Rs 3 lakh.

Finally, there's the strangest proposition of all: that both teams were aiming for the thrill of defeat rather than the agony of victory. Before the probe committee, Saeed said that he suspected the fix 'from the trend set by the Australians while bowling'. Sohail then told Qayyum that the 'match was played in a funny manner' and it 'looked that nobody was interested'. The judge attached no weight to their impressions, though Malik twittered in his recent merry meetings with *News of the World*: 'We were trying to get them to score runs against us and they wouldn't. We were trying to get ourselves out and they wouldn't get us out ... What a lot of aggro that was!' The game is now to be re-investigated by the ACB's new Perry Mason, Greg Melick. Who said one-day cricket was predictable?

Cursed by God

It was during this Singer Cup, it now hardly need be stated, that John the bookie befriended Mark Waugh and Shane Warne. And it was in its aftermath that Malik sought to ascertain if Waugh, Warne and

Tim May might like to play dead for dollars; Qayyum regarded the Australian evidence as 'rock-solid'.

Suspicion was rife behind the scenes during that Aussie tour of Pakistan. Intikhab Alam found a short-term solution to his problems: 'In the 1994 series against Australia, Asif Iqbal rang me up. Since I had my own doubts about him, I provided him with wrong information with the result that he never rang me up after that.' But when Pakistan then jetted off to South Africa, it wasn't so easy.

Evidence at the inquiry suggests that Pakistan's administrators were by now completely spooked. Intikhab told of being contacted in the dressing room during a Mandela Trophy match against New Zealand on 19 December 1994; it was Javed Burki saying that the match was fixed and that Pakistan was bound to lose. Intikhab protested that his team was 2/80 chasing 172, that defeat was impossible; luckily, he was right.

All that was missing at team meetings, meanwhile, was Jerry Springer with a microphone. According to Basit Ali: 'Aaqib Javed complained to Intikhab that bookmakers were coming to the rooms of the players and that he should take some steps. However, Ijaz protested and said he could not be asked not to see old friends. This led to an altercation between Aaqib and Ijaz and the meeting had to be dispersed.' Saeed Anwar, related Aamir Sohail, was particularly out of sorts: 'When I and Aaqib Javed were sitting with him in hotel, he said that he knew that he is not getting runs because he has taken money for fixing the match and that has come as a curse from God ... We told him that he should pray for forgiveness and pay some "kaffara".'

But the fissures really opened when Pakistan met the hosts in the Mandela Trophy finals on 10 and 12 January 1995. Twice Malik won the toss. Twice he elected to bat second, under South African lights that infamously offer as much illumination as a firefly in Carnegie Hall. Twice Pakistan was thumped.

In the first game at Newlands, Sohail hotfooted it to 71 off 74 balls out of 105, but was run out on the call of Ijaz Ahmed, who himself quickly got out as well. 'At that time we had to get runs at the rate of

4.5/5 runs per over,' Sohail complained to the inquiry. 'But surprisingly our batsmen played atrocious shots and got out.'

Told that players had been on the take, Intikhab insisted that they swear their honesty on a holy amulet before the second final at Wanderers. Malik consented to do so, but then didn't present himself until after the toss; by which time, Latif complained, there was no point. In the ensuing barney, according to Basit Ali, Ijaz sided with Malik and Waqar with Latif until Akram 'intervened and stopped the fight'.

After being stuffed by 157 runs, Intikhab thought a group hug in order. Latif told the commission that he was willing to embrace Malik, but that Malik 'did not respond positively'. As articles began appearing in Pakistani papers alleging corruption, Pakistan's president asked Burki to investigate, and an emissary was sent to South Africa: selector Salim Altaf. Like so many administrators before and since, he showed an absolutely fearless commitment to uncovering the truth: Latif claimed he was told to 'forget everything and concentrate on the game'. On flying home, Altaf had 'nothing to report'.

'The worst day in history'

In the weeks following the Mandela Trophy, Pakistan would have struggled to match it with an Amputees XI: they were rocked in a one-off Test at Johannesburg by 324 runs, then rolled at Harare by an innings and 95 runs. Curiously enough, Qayyum found little to suggest that the latter encounter, Zimbabwe's first Test victory, was bent: even Latif thought it an outcome simply of rock-bottom morale.

As Pakistan then fought back to overpower Zimbabwe over three Tests, publication of the Waugh/Warne allegations against Malik knocked the stuffing from the team once more. Intikhab had everyone swear on the Quran again, but Latif described the first one-day international against Zimbabwe on 22 February as 'the worst day in the history of the Pakistan cricket team'. Sohail and Ijaz had 'a fight … in the ground' and, with Pakistan needing a single to win with 2 balls left, Akram bunted a return catch to bring about a tie. It was after this game that Latif, having 'levelled allegations' to Burki in a Harare hotel, 'retired' and hopped on an airliner for home. Basit Ali joined him. Malik's reign of error was also shortly to end. Pakistan's cricket potentates, meanwhile, were united in their complete disagreement. Burki was 'absolutely sure' that Pakistani players were involved in match-fixing and recalled: 'Once Salim Malik was summoned to the presence of Arif Abassi and Zafar Altaf and was informed that he was involved in betting and match-fixing and cannot be allowed to play for Pakistan. He went off without refuting the allegations against him.' Zafar Altaf, however, continued to proclaim that his players were clean, telling the commission: 'I strongly refute the allegations against Wasim Akram, Salim Malik, Ijaz Ahmed and Saqlain Mushtaq. Their mothers can justly be proud of them.' And let's give their mums a bit of a break, for evidence since has hardly been definitive.

One game, which Qayyum didn't explore, might have resolved the mystery for good: the last in a three-match one-day international series between Pakistan and England in September 1996. Pakistan had lost the first two games, but looked a chance in this Trent Bridge fixture at 2/177 chasing 246 until quick wickets fell. Rashid Latif was purportedly told by his partner to get out, as a bet had been placed on England to sweep the series, and this colleague then set the example by throwing his hand away. Latif, however, stuck around.

If only Qayyum had been able to examine a fly on the wall of that Nottingham dressing room for, with two overs to go, Mike Atherton seemed to run Latif out with an underarm throw. But when the third umpire was booted up, it revealed that the stumps had been disturbed just before the ball's arrival by Alec Stewart's pads. Did his teammates cheer? Did they boo? Because if they were betting, they lost; Latif biffed his team to a two-wicket win. No fly, however, was available. And if anyone remembers how the team responded to that little green light, nobody asked them.

You must remember this

Not that memory proved entirely reliable at the inquiry either. Take Pakistan's second match against India in Toronto's Sahara Cup a fortnight after that match in England. Latif had actually been dropped, despite his recent heroics; hard luck, that. And according to Sohail, Akram left the field at one point, and he as vice-captain decided on a few overs of spin which captured a couple of wickets. Sohail testified: 'When Wasim Akram came back he was very angry and asked me what the hell you are doing. I said that I am trying to win the match and had taken two wickets for you.' Fishy, huh? But look it up. Pakistan won the game by two wickets thanks to a brilliant unbeaten 70 by guess-who: Salim Malik. Hey, don't confuse us!

Or consider another Pakistan versus India confrontation in the Wills Challenge Series a year later at Karachi when Saqlain Mushtaq, after giving away 13 in six overs, suddenly went for 33 off three, surrendering victory to the visitors. A red-hot fix, reckoned Pakistan's coach Haroon Rashid. No it wasn't, countered Saqlain. It was a late innings ball change that caused his inaccuracy: 'Since no old ball was available, a new ball was given after rubbing off its shine which created difficulties for the spinners to bowl and for that reason I could not contain the batsman.' *Wisden*, incidentally, backs Saqlain up.

You could go on but, after a while, all the matches start to blur into one: the Asia Cup, the Australasia Cup, the Titan Cup, the Coca-Cola Cup, the Akai Singer Champions Trophy, the Golden Jubilee Independence Cup, the Two-Flies-On-A-Wall Cup. For that's what this succession of rinky-dink slogfests for gaudy glassware amounts to: a cricketing Casino Royale at which it should not perhaps surprise us that chips slip off the green felt occasionally. Qayyum's conclusion is disarmingly humane and sane: 'that humans are fallible' and 'cricketers are only cricketers'. For it is, after all, other people who've decreed that Pakistan play almost 200 one-day internationals since 1994. Yes, they're the Incredible Exploding Cricket Team. But who lit the fuse?

Inside Sport July 2000

The King Commission
'The Gang Who Couldn't Shoot Straight'

AS CONSPIRATORS TO SUCH A GRAVE DECEPTION, Hansie Cronje and his cronies in South Africa's match-fixing fiasco are a disappointment. For a fortnight they have turned up before the commission convened by Judge Edwin King at Cape Town's Centre of the Book and, in sorry succession, sung like canaries. It's hardly Halderman and Colson, let alone Brutus and Cassius.

Even the testimony last Thursday of Cronje, the captain of the 'rainbow nation' so mesmerised by pots of gold, proved anti-climactic. Those expecting some sort of Afrikaner Moriarty at the centre of a web of deceit were confronted instead by the shambling shell of a man, mortified by his weakness for the 'lure of easy money', although still strangely oblivious to the extremity of his disgrace. He lurched between hand-wringing contrition and desperation for atonement, and a sort of schoolboy cheekiness, as though ready to cop a statutory six of the best before reverting to his former life.

Yet perhaps this is the most chilling aspect of the testimony heard so far: that it all appears to have happened so easily, so casually, so heedlessly of consequence. Corruption is usually like erosion, or sedimentation, a gradual process. But the evidence of opener Herschelle Gibbs regarding a one-day match against India three months ago, for instance, makes it seem a snap: 'Quite early on the morning of the match, Hansie came into my room with a huge grin. He said someone had phoned him to offer me $15,000 if I made less than 20. I said yes.' It sounds as routine as a batsman calling his partner for a single.

It was not. And while certain individuals in this affair are clearly more delinquent than others, everyone it has touched is culpable in some way. Thanks to decades of political and sporting isolation, South Africa's cricket community is unusually tight-knit and interconnected. Cronje attended the elite Grey College in Bloemfontein. His father Ewie had been a teacher there and later a leading cricket administrator who counted Ali Bacher, executive director of the United Cricket Board of South Africa, among his closest friends. Cronje's school coach Johan Volsteedt also coached his predecessor as South Africa's skipper, Kepler Wessels, and another player implicated in the affair, Nicky Boje. Cronje's brother Frans, a first-class cricketer, was best friend of Allan Donald, now South Africa's leading fast bowler. Their first captain at first-class level was Joubert Strydom, whose younger brother Pieter is among players to whom Cronje offered an inducement to fail. As teenagers, both Cronje and Gibbs received tuition from Englishman Bob Woolmer, then a retired county cricketer, later national coach.

It may have been this familiarity over decades that made the slide into malpractice so easy. Although it was almost two years since claims by Australian players had turned match-fixing into a global issue by the time the first serious approaches were made to Cronje, and through him to his men, it never occurred to anyone to report the inducements of bookmakers to their managers, to their administrators, to any of the three inquiries held in India, Australia and Pakistan during the last three years, or to the International Cricket Council.

Cronje, in one of his efforts at ingratiation last week while testifying, took the liberty of suggesting some remedies to the match-fixing problem: 'education' was one, quarantining of practice areas, hotels and changing rooms another. While intended to be helpful, they actually sounded ludicrous, as though dishonesty and cupidity could simply be legislated away. Despite all the disclosures about player corruption round the world, there is still no way to prevent it; if a player wants to sell his services, there will always be interested buyers. Perhaps, morally void as its habitués are, match-fixing is merely a logical outcome of the concept of sport as business and its leading participants as businessmen, auctioning themselves in the free market of entertainment.

Much of Cronje's mea culpa last week wasn't so much revelatory as confirmatory, with a few exceptions: that Pakistan's captain Salim Malik seemed to know that Cronje had been approached to rig the Mandela Trophy final in Cape Town on 10 January 1995; that India's captain Mohammed Azharuddin chaperoned him when he first met the mysterious Mukesh Gupta during the Kanpur Test of 8 to 12 December 1996. It was Gupta, Cronje has stated, who first offered him US$30,000 to induce a sub-standard South African performance in that Test, which he did not take to his team, then followed this up with an offer of US$200,000 to roll over in a limited-overs match at Mumbai two days later, which he did.

This game seems increasingly to have been a turning point in the saga. It was organised under bizarre circumstances – which, by the by, also demonstrate the complicity in the affair of cricket authorities, addicted to the staging of meaningless games. Originally designated as a benefit match for India's Mohinder Amarnath, the game was for no obvious reason reclassified as an official one-day international: the tour's eighth. The South Africans were peeved, Cronje particularly so. It suddenly became his hundredth one-day match; a landmark he'd been looking forward to achieving on home soil. In his 1997 team biography *Hansie and the Boys*, Rodney Hartman describes his subject's indignation: 'At no stage did any official bother to mention that the South African captain was playing his hundredth game. No official even shook his hand.'

The titillating novelty of the situation further prolonged discussion; wicket-keeper Dave Richardson likened the feeling to the *frisson* an unhappily married man might feel at the prospect of adultery. Even after opener Andrew Hudson, star batsman Daryll Cullinan and spinner Derek Crookes had spoken out sternly against acceptance, three players had lingered in their captain's room joking that their defeat might be worth more. In Pat Symcox's recollection, Cronje made a further call, eliciting the offer of an additional US$100,000, before saying that 'if the whole team was not involved he would not be comfortable about it' (Cronje says it was US$50,000). The game proceeded – a dreadful mess during which South Africa left the field for eighteen minutes in protest at missiles projected by an unruly crowd – and was lost anyway. Which perhaps tilted Cronje further from his natural axis: from this futile, thankless and ultimately irksome match, he returned more than usually empty-handed. When India arrived in South Africa for a reciprocal tour a week later, and Gupta approached him for information pertaining to team selection and a score at declaration, Cronje obliged and pocketed US$50,000.

There has been little dispute since the commencement of the crisis that Cronje, for all his private-schoolboy rectitude and Christian piety, has a covetous streak. Even Woolmer, a staunch supporter, has described him as 'money-driven', and described his 'whoops of joy and cries of despair' as the value of his share portfolio fluctuated. He might have met his wife Bertha Pretorious in a prayer group, but their lifestyle was hardly monastic; their R3.4 million mansion at Fancourt near the exclusive country club of which he is a member attests an enjoyment of the finer things in sporting life.

Three years after that first approach, team-mates have testified, Cronje went about his scheming like a greedy, even a vain, man. He himself has explained how a sports-betting executive, Marlon Aronstam, approached him on 17 January this year, the fourth evening of the Centurion Park Test against England. The Test had been ruined by rain, and only by the contrivance of a declaration could it be revived: Aronstam told Cronje that such a manoeuvre could dispel his image as a 'conservative and negative' captain, and by the way there was money in it, too. Greed and vanity were each rewarded, by R50,000 and a leather jacket.

What is truly striking about this admission is the degree of Cronje's recklessness; he accepted at face value an unsolicited offer from someone he had never met, and even advised Strydom to bet on the Test – in contravention of the United Cricket Board's evidently worthless code of conduct. He then bumped into a man whom he then knew only as Hamid – one Hamid 'Banjo' Cassim – and through him joined the circle of a London 'businessman', Sanjay Chawla.

It was this recklessness that brought Cronje undone. A Delhi police inspector happened to be monitoring Chawla's telephone traffic, including a cellphone given to Cronje. Cronje says that the phone seldom stopped ringing with offers and proposed scenarios. By the time he spoke on Thursday, Strydom, Lance Klusener, Jacques Kallis and Mark Boucher had already testified of Cronje's advice that money was available if South Africa agreed to come quietly in Tests beginning at Wankhede Stadium on 24 February and at Chinnaswamy Stadium on 2 March. Most seriously, Gibbs and Henry Williams had admitted to accepting US$15,000 each to follow orders in a one-day match at Nagpur on 19 March: Gibbs to make fewer than 20, Williams to concede more than 50 runs in ten overs.

Yet two features of the King inquiry testimony are still puzzling. The first is that not one of the conspiracies engineered by Cronje came off. Even when Strydom was encouraged to bet on the Centurion Park Test, he insists that it was on his own side: and not only did South Africa lose the game, but the wannabe gambler has claimed he couldn't get his money on anyway. Even when Gibbs accepted the offer to make less than 20, he made 73 'batting like a train'. Williams didn't get the chance to donate more than 50 runs, because he broke down after bowling eleven legal deliveries. The players could be lying. They could also be hoping to plead in mitigation that their performances were not influenced by cash offered or accepted. But on the basis of evidence so far, their activities look like those of the Gang Who Couldn't Shoot Straight.

Greed alone, too, can hardly account for Cronje's fall from grace, any more than Satan, at whom Cronje clutched in his first 'confession' on 2 May, like a drunk grabbing for a lamp-post. The weight of expectation on Cronje, which amid South Africa's process of 'Transformation' is immense and deeply politicised, seems to have impacted on him in a far more fundamental way.

The United Cricket Board was created in a merger of the official 'white' South African Cricket Union with the unofficial 'black and coloured' South African Cricket Board. Although Ali Bacher has proved a veritable Bismarck in reconciling past antagonists, strains have lately begun to show, specifically over the African National Congress's very public ambitions for a minimum number of 'players of colour' in national XIs. Sports minister Gconde Balfour has described himself as 'sick of lilywhite cricket teams'. The issue reached flashpoint eighteen months ago, after South Africa had made history by inflicting a 5–0 Test series defeat on the West Indies: literally a 'whitewash'. In his new *Woolmer on Cricket*, the former national coach recalls how his post-match reverie was disturbed by the sight of Bacher and Cronje in heated conversation: 'Hansie's face is like a book: when he is cross he has very frightening facial features. He smoulders ... and today he was upset.'

The *casus belli* was the squad for the forthcoming triangular one-day series: for the seven games, the selectors had named a squad bloated to seventeen by the inclusion of three 'players of colour', of which one, preferably two, would be picked for each fixture. At a team meeting the next day, Cronje announced his retirement, packed his bags and flew home. He likened his position to a chess player for whom someone else was moving the pieces.

Given various assurances about his autonomy, Cronje relented, and led South Africa in the one-day series. Just as well, contends Woolmer, otherwise we 'would have been down to very few men', for the 'support he had from team-mates was solid'. But South African cricket has ever since looked like an accident waiting to happen.

The national team has been exempted from the Transformation Monitoring Committee's three-year plan to achieve 'racial parity in all aspects of sport'. Bacher also spent much of August last year seeking – successfully – to dissuade Cronje from accepting a lucrative English coaching contract because the captain had 'an enormous amount to offer South Africa' and there was 'no-one better suited to captain South Africa at present'. Yet other events guaranteed to bring matters to breaking point again. Last August, UCB convenor of selectors Peter Pollock was displaced by Rushdi Magiet, formerly of the SACB. In February, UCB president Ray White was ousted on a no-confidence motion by another selector, Gerald Majola, formerly of the SACB. He was succeeded as president by Percy Sonn, formerly of the SACB. White went on his way denouncing the UCB as 'little more than the cricket organ of the ANC' and 'a cesspool of self-interest and politics'.

Did Cronje sense that his own time was limited? Was he apprehensive about the future of his country's cricket, so that he felt the need for present provision? Did he feel, as did the disgraced British politician Reginald Maudling when he explained his involvement in the Poulson corruption saga, in need of 'a little pot of money for my retirement'? If this was part of Cronje's motivation, it raises a gamut of issues that extend well beyond the cricket field. As it is, the affair already has some curly implications for post-apartheid South Africa: Gibbs and Williams were both 'players of colour' in the national XI.

Studying Cronje in the witness stand last week, it required some effort to recall that he was once regarded as among the most stoical and disciplined cricketers on the planet. That he was not impervious to temptation, that he was made to feel so insecure in his position, is telling. Sport has always been valued for the way it tests character. And at international level today, sport tests character in more ways than one can imagine.

The Bulletin June 2000

The Condon Report
The Dog in the Night-time

DEVOTEES OF SHERLOCK HOLMES will be familiar with 'the curious incident of the dog in the night-time', to which the great detective refers in the case of the stolen racehorse Silver Blaze. 'The dog did nothing in the night-time,' comment his puzzled followers. 'That,' replies Holmes at his oracular best, 'is the curious incident.'

Cricket, no stranger to curious incidents, has its own version of the dog in the night-time. In May 2001, the International Cricket Council published a report on the game's links with gambling. Sir Paul Condon, head of its Anti-Corruption Unit, lamented that 'corrupt practices and deliberate under-performance have permeated all aspects of the game', and described 'at least twenty years of corruption linked to betting on international cricket matches'. In doing so, it revealed a phenomenon that would have taxed even Sherlock Holmes' credulity: the dog that didn't bark for two decades.

Not that cricket's supranational governing body has been entirely inert. Thanks to its unfailing vigilance, sponsorship logos on players' shirts remain firmly within the bounds of propriety. Then there's that development program boldly pushing back cricket's frontiers, with the Eskimoes on track for Test status in 2150 AD. But will cricket last that long? Its new chief executive, Australian Malcolm Speed, found his in-tray full to bursting when he arrived a few months ago; even what might be considered the ICC's most significant achievements now appear mixed blessings at best. Cricketers are no more inhibited by the code of conduct than navvies by a swear box. Third-country umpires have simply brought about impartial incompetence. And above all, match-fixing abideth. Condon has described a 'core of players and others who continue to manipulate the results of matches or occurrences within matches for betting purposes'. The ICC devotes a page on its new website to 'the spirit of cricket'. But, surrounded by so many other pages devoted to enumerating that spirit's violations, it appears apologetic and archaic, the relic of a bygone age. Some would see the same as true of the ICC.

*

The ICC began with what for its time was a grand vision. When England, Australia and South Africa first met at Lord's in July 1909 under the auspices of the Imperial Cricket Conference, prime mover in proceedings was a South African plutocrat Sir Abe Bailey, whose immediate objective was a Triangular Tournament to be held in England. History has been unkind to that ambition. The idea of inter-locking three-Test series involving three countries on the same soil was far-sighted: unfortunately for Bailey, so far-sighted that it ulti-mately proved a miserable failure. It is fair to say that the ICC has never repeated the mistake of being ahead of its time.

The Imperial Cricket Conference began meeting annually at Lord's seventy-two years ago and, while it gradually took in the West Indies, India, New Zealand and Pakistan, remained essentially an adjunct of the Marylebone Cricket Club: chairman and secretary of the ICC were concurrent offices respectively of the president and secretary of Marylebone. Not everything went swimmingly. Members had to con-tend with the 'chucking' pandemic at their meeting in July 1960, the conference being the only forum suitable for discussions about cor-rective action. The constitution of the conference then required that South Africa, having quit the Commonwealth to become a republic, be expelled. Otherwise there was little done, because there was little to be done. Even as international cricket expanded, its management remained as simple as a post office box and a filing cabinet at Lord's. At the end of the usual two-day meeting, Marylebone's secretary would drop off a short communiqué to various press agencies, which *Wisden* would publish somewhere deep in its least-consulted recesses.

Nowadays we can all have a chuckle about this, the old buffers at Lord's and their Land That Time Forgot, caught out by the crumbling of the British empire, and sounding increasingly like the corrupt police chief who closes down Rick's Café Americaine in *Casablanca*: 'I am shocked, *shocked*, to find that there is politics going on in this establishment!' But the old ways were not merely a matter of paternalism and condescension on the part of fuddy-duddy imperialists: as well as being a geopolitical story, the ICC is a regulatory story, explicitly the eclipse of an old system reliant on unspoken codes of behaviour rather than on rules and statutes.

In terms of its regulation, international cricket has much in common with another great British institution: the City of London. Both come from traditions of informal checks and balances, based on the assumption that their constituents knew how to behave, and where they didn't that a tap on the shoulder and a talking-to usually did the trick. The virtue of such models of management is their simplicity, informal understandings taking the place of legislative rigidities. Their weakness is that they are robust only up to a certain point, and that determined defiance sets them at nought.

One of the decisive moments in the history of the City, for example, was the so-called Aluminium War in late 1958. Orchestrated by the brilliant financier Siegmund Warburg, the bid by Tube Investments and the American Reynolds brothers for British Aluminium was the Square Mile's first significant hostile takeover, and was fought with extraordinary vehemence. British Aluminium's chairman, Viscount Portal of Hungerford, spoke of having to 'save British Aluminium for civilisation', and sought a 'white knight' in the shape of Alcoa. But despite opposition from every leading City institution, and intercessions by the governor of the Bank of England and the chancellor of the exchequer, Warburg refused to knuckle under; he went into the market, and bought the company from beneath the obdurate Viscount. 'The establishment had lost,' William Davis explains in his *Merger Mania*, 'and, worse, it had made itself look extremely silly.'

The City was never quite the same. A partner at a rival bank told Warburg: 'No company director whose shares are publicly quoted can sleep well from now on.' And it's chiefly as a result of the Aluminium War that the first rules on the conduct of takeover bids were set down: something unthinkable before it. Viscount Portal, meanwhile, retreated to his other job: he was at the time president of the Marylebone Cricket Club and *ex officio* chairman of the ICC. At least there, one imagines, he could feel that chaps knew how to behave.

In May 1977, the ICC faced its equivalent of the Aluminium War: a hostile takeover of international cricket by a wealthy outsider, Kerry Packer, again fought with extraordinary vehemence. Like British Aluminium, cricket's establishment went looking for a 'white knight': England's Test and County Cricket Board hopped into bed with Cornhill Insurance. 'It was difficult to avoid the temptation of thinking that Packer would simply go away,' recalled ICC secretary Jack Bailey. 'It was as if he didn't understand that "authorised cricket" held the world copyright on the game and its organisation.' But like Warburg, Packer was not for turning. He too went into the market, as it were, and bought the company from under the ICC, by recruiting most of the world's best players to his inchoate World Series Cricket. The gruelling case that Packer forced the ICC and TCCB to fight before Justice Slade of the High Court demonstrated that informal understandings, however honoured in tradition, counted for little.

The ICC's constituents were luckier than the board of British Aluminium. Packer had no aspirations to own international cricket outright: his objective throughout had been the broadcast rights for cricket in Australia, and this ultimately seemed a small price to pay for the restoration of peace. But the animal spirits of the market, once liberated, were irrepressible. The ICC appears in dire straits now, yet the situation in the 1980s was perhaps even worse. Players went toe-to-toe with umpires on the subcontinent, threatened to abandon tours there, queued for handsome testimonials at the new stadium in Sharjah and for rebel tours to South Africa. The ICC, however, had begun to perfect its 'dog in the night-time' impression: from Lord's, there wasn't a peep until January 1989, when a resolution was passed ruling South Africa officially off-limits to international players. Mike Gatting's rebel team showed how much a deterrent this was by signing to go to South Africa six months later.

It was also in this period, Sir Paul Condon believes, that cricketers enjoyed their first flirtations with bookmaking interests: India's Central Bureau of Investigation ascribes the rise of cricket gambling to the resumption of competition between India and Pakistan in 1978. In hindsight, coincident with the professional revolution wrought by Kerry Packer, the timing for match-fixing was perfect: what is it but professionalism *in extremis*? In his landmark High Court judgement, Justice Slade set the rights of cricketers in stone: 'A professional cricketer needs to make his living as much as any man.' But what if the highest bidder for your services is a bookmaker? The only thing to prevent you taking the money is conscience, which as H.L. Mencken reminded us is merely 'the inner voice that warns us someone might be looking'. And in the 1980s, nobody was looking and nobody wished to.

*

Considering the experience of all the king's horses and all the king's men, it is remarkable how often institutions try reassembling their regulatory Humpty Dumptys. Even after the Aluminium War, the City was resistant to the heavy hand of statutory authority, and shrank from the idea of a police-and-punish regulator like Wall Street's Securities and Exchange Commission. There was no need for that, it was felt, providing everyone played the game. But not everyone did.

Nine years after the first Takeover Code was drafted in 1959, it was superseded by another stricter version, this time policed by a Takeover Panel. But that soon started fraying at the edges as well. For the next two decades, in fact, City regulation would continue to lag behind market developments, as though hoping at some stage for a collective seizure of morality to restore the old understandings. With deregulation in October 1986 came the Financial Services Act, and a bevy of new administrative bodies including the Securities and Investment Board and the Investment Managers Regulatory Organisation. But the Takeover Panel resisted a House of Lords motion that it be vested with statutory powers, even that it be subject to judicial review. Robert

Alexander, QC, who became chairman in May 1987, believed that the panel was 'generally effective in monitoring the conduct of takeover bids'. Another little coincidence: Alexander, now Lord Alexander of Weedon, was the last president of Marylebone Cricket Club.

Cricket took a cautious step forward in July 1989 when Marylebone surrendered its role in the management of international cricket. But despite taking the name International Cricket Council, the members who appointed Lord Cowdrey as chairman with an administrative staff of one can scarcely be accused of revolution. As with successive City Codes, it was an attempt to move into the future without abandoning the past, underpinned by a voting structure that staved off full democracy by privileging the entitlements of 'foundation members' England and Australia. Chaps could still be trusted to behave – but only up to a certain point.

That point was reached in February 1993 when India wrested the next World Cup from England with the assistance of Pakistan, Sri Lanka and the associate members. That thirteen-hour shakedown – the first significant bust-up at the ICC twixt East and West – resulted in another shake-up. The council would obtain its first full-time secretariat: a non-English chairman in West Indian Sir Clyde Walcott and a non-English chief executive in Australian David Richards. Again, however, this changed the accents at Lord's rather than the accent of cricket: the new ICC began with two administrative staff. Walcott was never other than a *roi faineant*, while Richards demonstrated a Bismarckian ability to remain silent in seven languages.

This is curious. On one hand, the ICC *was* moving to codify what had previously been taken for granted. After the excesses of the 1980s, there was botheration about player behaviour: in June 1990 and January 1991, members agreed in principle to third-country umpires and to a retinue of referees to enforce a code of conduct, essentially an attempt to preserve cricket's spirit by a balance and a check. But what held for the players was not thought to hold for the game's governors. International cricket, it was felt, had no need for a centralised, overarching administration: it could run on the basis of putting on your best suit and going to Lord's a couple of times a year.

Where self-regulation is preferred to legislation, there is usually a public justification and a private reality. Publicly, it is stated that everyone works better in an environment free of bureaucratic impediment and governed simply by the unstated rules of mutual reciprocity. Privately, it is also acknowledged that everyone wants the freedom to do as they please. That was the City's experience. It has also been cricket's. The ICC remained weak because its members liked it that way.

In one respect, the ICC was actually less powerful in the 1990s than in the 1960s. In olden times, the ICC had at least signed off on international schedules. But with the increasing popularity of limited-overs cricket, arrangements grew increasingly *laissez faire*: nothing prevented boards of control agreeing bilaterally to gimcrack one-day tournaments wherever an entrepreneur could erect some stumps twenty-two yards apart and plug in a few television cameras. At the time of the World Cup punch-up in 1993, the world had witnessed 800 one-day internationals in twenty-two years. That number doubled in the next seven.

The effect is now clear: as Condon puts it, 'an environment in which it was possible to watch and bet on cricket almost every day of the year'. But of this beanfeast for bookies and bribefest for bent batsmen and bowlers, the ICC was no more than an onlooker. Richards was in Pakistan in October 1994 when Salim Malik bid for a bad day's work from Shane Warne, Tim May and Mark Waugh. Richards and Walcott were then informed that the Australian Cricket Board had fined Warne and Waugh for accepting money from a bookie for information, and forwarded their sworn statements. But they did nothing. Technically, the problem was constitutional. The ICC's code of conduct prohibited 'use and distribution of illegal drugs' (which had never been an issue), but not use and distribution of illegal cash (which now was). Its handicap, though, was more philosophical. Consider two questions. Would the ICC have been wise to seek the power to hold an inquiry from its members, and would it have obtained it? The answers, respectively, are probably yes and almost certainly no. Just as governments rolling in gaming

revenue shrink from addressing the social cost of casinos, the boards
of cricket authorities waxing fat on television rights and sponsorships
tend to think that the more rinky-dink cricket the better.

*

So the crucible of corruption was heated. In the days of the *Pax
Britannica*, the ICC had been like a club: cosy, gentlemanly and
inward-looking. It still resembled a club during the 1990s but one
where nobody liked anybody else much, thanks to post-colonial fric-
tions and naked self-interest. Which brings us to Jagmohan Dalmiya.

When Dalmiya crusaded for the ICC chairmanship in July 1996,
using as his election platform the rich pickings of the recent World
Cup, it was hard to resist the logic of his appointment. India was a
growth economy: it inhaled cricket and exhaled money. Dalmiya's
campaign was crass, and the legal wrangle that accompanied it as
unedifying as watching the last US election fought down to the last
chad. But the compromise involving yet another restructure of the
ICC in June 1997 that allowed Dalmiya the new position of president
appeared workable.

In fact, Dalmiya lived up to La Rochefoucauld's maxim that it is
easier to appear worthy of a position one does not hold than of the
office one fills. He made money, though we can't say how much
because the ICC keeps such internal affairs as confidential as the
Knights of the Templar. He had a vision of globalisation, though it
was about as sophisticated as throwing darts at a map of the world.
And even if not actually corrupt himself, he was a do-nothing leader
on the most important issue of all. When the corruption story began
breaking in India as he was about to commence his ICC presidency,
he scoffed; these were attempts to 'malign the name of Indian cricket
for the sake of a juicy story'. He stated unequivocally: 'I don't think
the players are involved at all, and if it is proved it will be the shock
of my life.'

Dalmiya's tune had changed somewhat by the ICC's meeting in
Christchurch in January 1999 – he stated that 'everything would be
done to probe this sordid affair'. But the song remained essentially

the same. The powers of the Code of Conduct Commission set up as a result of that summit were restricted to reviewing investigations made by other countries. Its first act was to endorse India's Chandrachud Inquiry, which now appears to have contained more dissembling than Bill Clinton's grand jury testimony. In the meantime, Dalmiya specialised in events of the *panem et circenses* sort. And while Dalmiya was acclaimed for taking the Wills International Cup in October 1998 and ICC Cricket Week in April 2000 to Dhaka, one wondered at the mentality of an ICC president who indulged in such empty gestures as obtaining the patronage of the United Nations for the latter festival. 'Cricket is renowned as "the gentleman's game" where all matches are not only governed by rules but also by the spirit of cricket,' he sermonised. 'The phrase "it's not cricket" is known over the world and emphasises fair play.' As we say in Australia, which game was he watching?

*

In September 1986, the City of London received a very loud wake-up call. In the course of his confessions to the American Securities and Exchange Commission, insider trader Ivan Boesky admitted that he had rigged the price of Distillers PLC on behalf of its suitor Guinness. One of the City's greatest scandals came to light – Guinness's chief Ernest Saunders was sentenced to five years' jail for what the judge called 'dishonesty on a massive scale' – although thanks only to the chance discovery of another regulatory body.

Cricket was the same. Even as Dalmiya was wittering on about 'the gentleman's game', policemen in his country had the solidest proof yet that it was not. Their interception of Hansie Cronje's telephone traffic with Sanjay Chawla – a matter of pure chance, and owing nothing to the game's authorities – presented cricket with a bullet hole, smoking gun and fingerprints. Sad to say, that luck was not best exploited.

Sometimes, you suspect it's fortunate that cricket administrators only run cricket: if they had been prosecuting Timothy McVeigh, he would now not only be free but hosting his own chat show. Although

Cronje attracted a life ban as a result of his testimony before the King Commission, his lawyers were still able to bring the inquiry to a premature conclusion by mooting a legal challenge to its constitutionality, while South Africa's United Cricket Board meted out to his co-conspirators Herschelle Gibbs and Henry Williams not so much sentences as syllables. Once again, the ICC stood mocked by one of its own.

As observed, the ICC is exactly as powerful and influential as its members have made it. And, fond as they are of investing their meetings with the gravity of a G7 summit, they usually seem happy for the ICC to remain as ceremonial as the plastic bride and groom atop a wedding cake. When the Anti-Corruption Unit tried to follow up Pakistan's Qayyum Inquiry early this year, Pakistan Cricket Board president Tauqir Zia inadvertently put it as well as anyone: 'The ICC is nobody to tell the PCB that these boys cannot play. My stand is very clear. If we want them to play, they will play.' In this realm, to borrow Yeats' words, the best lack all conviction and the worst are full of passionate intensity.

Nor does this apply only to match-fixing: the ICC's recent history contains many other classics of backsliding. One was perpetrated by the Australian Cricket Board when Allan Border's team met South Africa at Johannesburg in February 1994, and Shane Warne and Merv Hughes earned themselves appointments with the ICC beak for what we might describe euphemistically as letting the heat of the moment impair their judgement. It was exactly these incidents for which the ICC's code of conduct had been devised, and referee Donald Carr docked the players accordingly: just over A$400 as it happens, although there were mitigations in terms of the earbashing the Aussies had received from a merciless and menacing crowd.

In Australia there was a furore. The incidents were replayed endlessly and talkback radio hummed for days; the ACB got a drubbing. So for the sake of its own reputation, it slapped Warne and Hughes with fines of A$4000 each, over and above the levies of the ICC. Which is not to say that Warne and Hughes did not deserve punishment, nor even that Carr did not let them off lightly. It is to say that by trying its players twice for the same offence as a public relations gesture, the ACB gratuitously undermined the authority of the ICC.

An even more offensive instance of hypocrisy was then perpetrated by the Board of Control for Cricket in Sri Lanka in the wake of the no-balling of Muttiah Muralitharan by Darrell Hair during the Boxing Day Test of 1995. You may remember that Sri Lanka, India and Pakistan insisted that the ICC constitute an Advisory Panel on Illegal Deliveries to arbitrate on such disputes, where Murali passed muster (although he was never 'cleared': the Laws still leave the final decision in the umpire's hands).

There have been darker days in cricket than the one on which Murali was then called during an Adelaide one-day international in January 1999, though not many. If Arjuna Ranatunga's fifteen-minute huff while he remonstrated with umpire Ross Emerson and consulted his Board president did not 'bring the game into disrepute', then the code of conduct isn't worth the *Wisden* it's printed in. Again, however, the real transgression here was obscured. It wasn't Murali's action, Emerson's actions or even Ranatunga's reactions; it was the response of the BCCSL which, when Ranatunga faced referee Peter van der Merwe in Perth a few days later, equipped him with more lawyers than O.J. Simpson. Their contention: that the ICC had no right to hear the case. So while the ICC was the only possible forum for deciding Murali's fairness, it should have no say in deciding Ranatunga's. A paradox, a paradox, a most ingenious paradox.

The issue raised by Ranatunga's lawyers about the ICC's code of conduct was important: referees playing the role of investigator, prosecutor, judge and jury must pose questions about natural justice. Yet was not the BCCSL part of the body that had created this code? Had the BCCSL not nominated referees to enforce it? And if the BCCSL sincerely held its natural justice concerns, why had they not been raised before? Again, here was a board prepared to support the ICC when it suited, but poised for petulance when it did not.

There was an almost exact sequel a year later when the ICC Advisory Panel ruled just that Shoaib Akhtar's action merited closer scrutiny. Notwithstanding their insistence on the panel's creation, and that they had never advocated an appeal mechanism, the Pakistan Cricket Board went straight to Dalmiya demanding the decision be rescinded. The ICC president promptly remitted Shoaib on the grounds that only his bouncer, which would perforce be no-balled in limited-overs cricket, was under examination.

Here once more the issue became blurred. It was not a matter here of whether Shoaib transgressed. It wasn't even a question of Dalmiya's twisted logic. It was an issue of whether a process that the ICC had instituted, and which the country now protesting had actually demanded, should be allowed to run its course. The ICC isn't merely a chain as strong as its weakest link; it is a chain where the links have competed for that honour.

*

We may be moving on. There is nothing like a good series of scandals to precipitate regulatory reform. It happened in the City. To use the title of David Kynaston's superb new history of London as a financial centre since the Second World War, it is 'a club no more'. In June 2000, the Financial Services and Markets Bill collapsed nine separate City regulators into a powerful single body, the Financial Services Authority; although the Takeover Panel survives, it has been relegated to a subordinate role. In addition, under the European Union's Takeover Directive, the Takeover Code finally has a statutory basis: the days when squabbles were resolved by lunch rather than litigation are long gone.

The ICC is moving down the same route. It now has an executive board and a new CEO with a growing bureaucracy. Since the US$550 million obtained for television rights to the next two World Cups from World Sports Group/News Corporation, there is really no excuse for there not to be a responsible and properly resourced global management body for cricket. It has even, belatedly, secured a measure of control over the scheduling of international competition: its International Test Championship, while flawed, is a vast improvement on the prevailing adhocracy. But an arduous road lies ahead; Condon's candid view was that the ICC 'does not yet provide an infrastructure to meet the financial and governance requirements of the modern game'. And if it is to be effective, the ICC will have to ford one other frontier, perhaps the greatest of all. We have talked about the ineffectiveness of morality without statute. But statute without morality is equally nugatory.

The ICC has contributed by neglect and omission to the crisis in which cricket finds itself. Yet the blame cannot be heaped entirely at its door. Ultimately, it is cricketers who have brought discredit on their game. Administrators merely let them. The ICC has accepted several recommendations from the Condon report that, while perhaps useful in the interim, simply should not be necessary. It intends what it describes as a training program 'to raise awareness of the risks of corruption in cricket and the methods used to entice players and others into malpractice' which will 'emphasise the resolve of the ICC to deal with the problem and to punish wrongdoers'. It also wants each Test country to appoint a 'security manager' for teams during international competition, their tasks including 'providing advice and action in relation to the security of players, officials and venues' and 'preventing and detecting improper approaches to players on tour'.

That either of these proposals should be considered is an indictment of professional sporting morality. Is the only way to ensure that cricketers do not fall among thieves to 'scare them straight', and quarantine them from importunate strangers with mobile phones and fat chequebooks? Is it truly the case that unless extensively trained and supervised, cricketers are incapable of knowing that it is a bad thing to sell your wicket, or donate runs, or throw away a match, for money? If so, then it is arguable that the whole effort is futile: it may reduce the incidence of corruption, but never eradicate it. Fear of punishment has never stopped crime. The tightest security cordon can be breached. And frankly, even now, if a player wishes to sell his services, he will always find interested buyers.

More constructively, Condon hinted at a new direction for the game. Cricket, he contended, would have to pay more attention to the players: they were, he thought, 'not sufficiently involved in the administration of the game and ownership of the problems'. At the same time, the players will have to pay more attention to cricket. Without their commitment, all the policies, prohibitions, inquiries and investigators won't make a jot of difference. This is now more than a matter of conscience: if players are so oblivious to the game's welfare that they will tarry with bookmakers, and contentedly keep important information to themselves, this may be the last generation that takes cricket seriously. The ICC's credibility has been questioned freely throughout this affair. But how credible are the players? Three international captains – Hansie Cronje, Mohammed Azharuddin, Salim Malik – have been banned for life. Three others who have led their country – Shane Warne, Waqar Younis, Wasim Akram – have been fined. Five more – Arjuna Ranatunga, Aravinda de Silva, Alec Stewart, Brian Lara and Martin Crowe – have or are being investigated. Is this a case of a few bad apples, or a whole barrelful?

Those who pillory the ICC for its inaction on match-fixing then, ignore the sorry fact that it should never have needed to take action in the first place. The reason that the dog did not bark in the night in the case of Silver Blaze – which gives Sherlock Holmes the clue he needs to solve the case – is because it recognised the intruder and assumed his intentions honourable. The dog was merely mistaken. It was the intruder up to no good.

The New Ball Volume 6 2001

Match-fixing
Cronje's Fall

'THE EVIL THAT MEN DO LIVES AFTER THEM,' intones Mark Antony over the body of Julius Caesar in Shakespeare's play. 'The good is oft interred with their bones.' It would be an exaggeration to classify Hansie Cronje as an evil man – he was, at worst, amoral, greedy and vain – but his is to cricket an altogether malign legacy, and one that the tragedy of his death will not efface.

The fact is we now exist in the post-Cronje world. Every mouthing about 'the spirit of cricket' and 'the noble game' is inscribed with irony. Every unexpected international score-line occasions a musing about who might have been enriched. Every international team will shortly be chaperoned by its own security manager – tacit acknowledgement that everyone is venal and nobody beyond price. In the International Cricket Council's unending stream of ordnances, even those unrelated to match-fixing, one detects a determination to establish that 'strong government' for cricket whose absence has been blamed for Cronje's delinquency.

There's been little contemplation, however, of what Cronje's corruptibility might suggest about cricket in particular and professional sport in general. Most tend to treat his fall as part of a continuum of controversy with us since rascally Dr Grace, the last chapter in that bumper book of cricket scandals. But ethical transgression in sport has customarily been concerned with competitive advantage; Cronje and his crooked contemporaries perverted cricket in a way it had not been since the Lord's bookmakers followed wretched William Lambert into exile 180 years ago.

The Cronje case fits better on a continuum of commercialisation. It is forty years since the law recognised sport as a trade in *Eastham versus Newcastle United*, where Justice Wilberforce enshrined an athlete's right to act against an 'association of employers whose rules and regulations place an unjustifiable restraint on his liberty of employment'; it is twenty-five years since Justice Slade's landmark judgement in *Greig versus Insole*, which explicitly recognised that 'a professional cricketer needs to make his living as much as any man'. Match-fixing was merely the ultimate expression of sport as 'employment', and of the right to sell one's prowess to the highest bidder. Many were incredulous that Cronje – so earnest, so intense – should have fallen among thieves; yet what better candidate for corruption than one whose cricket was so drab, cold and joyless?

In fact, Cronje reminds us how pronounced are the tensions when an athlete is trained to think in terms of career, where what began as play becomes work, where loyalty to country, team, board, fans and game competes with loyalty to self. It's easy to forget that, in historical terms, professional sport is a very recent development, and that, strictly speaking, its sustainability remains unproven. On one level, public response to Cronje's disgrace was reassuring; it challenged the dangerous trend to thinking of sport and entertainment as indistinguishable. If sport was merely another branch of the entertainment industry, then it could be argued that Cronje let no-one down. Spectators still got to watch a game of cricket, and such tampering with outcomes as took place did so at the margins: it was no more disappointing than, say, an Elton John concert without a third encore. But, as we know, it felt otherwise. There is actually a solid residue in sport of the amateur ethos often assumed to have been completely overthrown. Sport contains values that transcend the commercial; otherwise, it's not sport.

Cronje's betrayal broke sport's most solemn trust. In his classic of sociology, *Fights, Games and Debates*, Anatol Rapoport pointed out a precondition of any game is 'the assumption of similarity' about our rival: 'Playing a game is impossible unless one can assume that one's opponent intends to win if he can, and that, in trying to achieve this, he will be influenced by similar considerations, and have in mind the same kinds of strategies.' Cricket the Cronje way was another thing entirely.

An economic historian reviewing Cronje's corruption would probably detect a degree of unavoidability about the whole affair. It's instructive to consider its parallels with that benchmark case of twentieth-century sports bribery, the 1919 Black Sox scandal, when the infamous Arnold Rothstein paid eight Chicago White Sox players to gift the World Series to the Cincinnati Cubs.

Firstly, there are similarities between baseball then and cricket now in their easy assimilation of gambling culture. By 1919, baseball was served by a huge legal gambling apparatus permitting bets on a host of variables, to which authorities turned a blind eye on the grounds that they broadened baseball's appeal. Cricket, of course, has a long acquaintance with gambling, and legitimate betting on the game is now widespread, pervasive, and still sanctioned by administrators as innocent fun. I suspect this is more fundamental to the Cronje case than has been recognised. It's been forgotten how Cronje's slide into malpractice was abetted by a representative of the legal gaming industry, Marlon Aronstam of South Africa's NSI, for whom Cronje contrived the outcome of the Centurion Park Test thirty months ago.

In both baseball and cricket, market forces then applied. When the US entered the First World War in April 1917, the government closed racecourses, sending footloose gamblers and bookmakers flocking to baseball with a fresh supply of 'hot money'. At a time when baseballers had just accepted a hefty pay-cut, reflecting wartime austerities, corruption was almost inevitable. The scenario was similar in India eighty years later. Its only legal form of gambling was horseracing, extensively regulated and taxed. In response, gamblers and bookies had shifted steadily to cricket, unregulated and untaxed. This at a time when some cricketers were doing rather better from the game than others – South Africans, in particular, were in relative terms one of the poorer-paid international teams, thanks to the rand's decline against the US dollar. An established gambling culture, a stampede of speculators, players with shrinking income: in baseball and cricket, these have proven a toxic admixture.

A student of the Black Sox and Cronje cases would, however, be struck by their antithetical outcomes. It is forgotten today that the eight baseballers held to account eighty years ago were, for various reasons, acquitted of conspiracy at the subsequent Grand Jury hearings. It was the baseball commissioner who the day after their acquittal banned the players for life – even those who had refused to accept money and merely been present at meetings where the Rothstein inducements had been discussed.

In contrast, Simon Wilde notes in his recent book *Caught*: 'No South African players were punished in connection with match-fixing, despite most of the team sitting on information relating to the issue for years.' Some, in fact, have done rather well. Herschelle Gibbs has returned from his wilderness minutes. Pat Symcox and Andrew Hudson have been appointed to a high-profile committee charged with a root-and-branch review of their country's cricket. Dave Richardson has made a giddying ascent to running the ICC's cricket operations department. If not offenders in the Cronje class, they were part of the climate of silence and neglect in which match-fixing prospered.

In this whole affair, however, shame has been in short supply. No player has stepped forward and said: 'Although I didn't accept money, I knew that others did, and kept quiet about it out of a mistaken sense of loyalty to other cricketers that blinded me to my duty to cricket.' Nor has any administrator hastened to concede: 'I may not have known details of match-fixing but I consciously made no effort to find them out, forgetting that the running of a sport is about more than its short-term profitability.' One might have expected over the last two years some debate about values, some sort of moral reflation, some recognition that cricket's integrity is everyone's responsibility. Instead, the responsibility has essentially been outsourced to Sir Paul Condon.

Thus the new, post-Cronje spirit of cricket: be ethical – or else. It's as though cricketers shouldn't be expected to exercise moral judgement; that they should be free to concentrate on which nightclub to patronise and which brand of sunglasses to endorse without having to worry about pesky distractions like distinguishing right from wrong. Sooner or later, it would seem, every aspect of behaviour in cricket, on field and off, will be regulated by statute and policed by official supervisors, rather than a matter for individual conscience. It would be unkind to suggest that this is inevitably the result when a lawyer and a policeman take charge of anything; after all, a strong ICC might be desirable for other reasons. But will any of this waken cynical cricketers and sleazy administrators to their responsibilities for the game's welfare? Or will it just make them more careful not to be caught?

Wisden Cricket Monthly July 2002

The ICC

What Is the Point of Cricket Administrators?

BILL O'REILLY CALLED THEM THE COAT-CARRIERS. Administrators, that is. Oh, they're necessary all right. After all, before any game of cricket is played, it must first be organised. Yet how often we bemoan their shortcomings – those failures of political will in the face of the bureaucratic won't.

The history of the cricket administration in this country is populated by buffers and blimps, some benevolent, some not, most of them well-meaning mediocrities interspersed with a few downright bullies. Many endured for decades. In discussing the attrition among Test captains, Richie Benaud once used the contrast of the eternal administrator: 'There may occasionally be a case of a minor official getting the chop but it is rare indeed for the more established ones to feel the touch of cold steel on the neck.' He added, in that endlessly imitable deadpan: 'It has always intrigued me that this should be so.'

As for the question of global cricket government, it might until recently have elicited as a response Gandhi's opinion of Western civilisation: 'It would be a good idea.' But lately, especially since the ascension of Malcolm Speed to the role of CEO at the International Cricket Council, inertia has been replaced by ertia, and the house of Lord's has re-established its dominion. The 2003 World Cup in South Africa, if not its finest hour, was by some margin its most conspicuous: its corporate symbol was to be seen everywhere; its corporate clout was felt in the unlikeliest places. When spectators had Coca-Cola confiscated at Durban in order to preserve the terms of a sponsorship agreement, cricket administration reached a new level of something – it's just not quite clear what.

Let me confess some fellow feeling for Mr Speed. I, too, am a cricket administrator. At time of writing, I'm vice-president of my Melbourne club team, the Yarras, not to mention its chairman of selectors, newsletter editor, trivia night quizmaster and reigning karaoke champion. I collect subs, pay the curator, man the bar and make sure the tab does not exceed consolidated revenue – like so many others round Australia, and the world, who keep their clubs just the right side of the brink of extinction.

At one time, of course, all cricket administration was of the kind in which I am involved. When Richard Nyren ran the Bat & Ball Inn in Hambledon's heyday, you can bet he cursed unforeseen unavailabilities, importunate creditors and the dearth of good tunes on karaoke machines as much as I do now. Yet there is much to be said for that simplicity of objective. Flags and honours add some glister. But, for a club, success translates simply as prolongation of existence.

The purpose of the modern administrator of international cricket is not so clear. To be fair, Speed has eliminated much of the doubt that used to surround what the ICC was doing; the ambiguity is now related to why it is doing it. Where once there was a post office box and filing cabinet in NW8 is now a sizeable freestanding bureaucracy. But to borrow the famous question of one of Mr Speed's legal antecedents: 'Cui bono?' To whose benefit?

It is obvious that the responsibilities of a sporting authority now extend beyond simply ensuring that its sport survives. But bottom-line imperatives aren't the be-all and end-all either. The management of sport requires the application of some business principles, but no set of straightforward commercial performance indicators suffices. There are no recalcitrant investors or equities analysts out there saying that a cricket board need achieve the same ROE or per capita productivity ratios as, say, that particularly venturesome badminton association. But nor do cricket boards have the option of deciding that their game's prospects are limited and diversifying into cribbage or polo. They have all the advantages – and susceptibilities – of being stable, slow-growth monopolies.

National cricket bodies can at least seek affirmation through the performance of their Test and one-day teams. If these are prospering, the authorities are generally assumed to be doing the right thing, even though the connection of the two is essentially unprovable. The rise of the West Indies as a cricket power was not the achievement of a strategic fiat of the West Indian Cricket Board of Control. The flattering light that Australia's achievements cast on the Australian Cricket Board may or may not be justified.

What of the ICC? Harold Macmillan once described the life of a foreign secretary as being 'forever poised between a cliché and an indiscretion'. As a multinational body with a few refractory members, the ICC has tended to operate along similar lines, with cliché usually in the ascendant. But it might be gainful if we worked out what the body stood for, using its most recently stated aims.

Consider, firstly, the ICC's 'mission statement' as published in its latest annual report: 'As the international governing body for cricket, the International Cricket Council will lead by promoting the game as a global sport, protecting the spirit of cricket and optimising commercial opportunities for the benefit of the game.' On the ICC website, this is further explicated by a note: the ICC's 'four key responsibilities' are delineated as 'leadership, promoting the globalisation of cricket, maintaining and enhancing the traditional spirit of the game, and ensuring the commercial prosperity of cricket'. At first, this sounds like the pointless duplication into which corporate-speak so often lapses. Listen a little harder and you'll detect a slight creak. 'Leadership' is inherently singular, 'globalisation' optimally plural; the 'spirit of cricket' seems to come under stress when stakes are high, yet the ICC is simultaneously committed to making those stakes higher still. These stresses aren't irreconcilable, but they are inescapable.

The ICC has also published a list of values – eleven of them, which if intentional is charming in its way – to be upheld for the benefit of 'stakeholders'. Some resemble what we might colloquially call 'Dalmiya clauses': boilerplate statements for the sake of peaceful coexistence between Mr Speed and the mercurial Indian board secretary Jagmohan Dalmiya. One value promises to work in a 'united and cohesive manner'; another to foster 'trust and respect throughout the ICC extended family'; a third further defines respect as involving 'recognition of authority and respect for cultural diversity'. In other words, we promise to be ever so sensitive ... providing everyone toes the line.

Then there's that quaint and fusty old spirit of cricket, which despite abundant offences against it so stubbornly refuses to die. Under the heading 'Tradition', the ICC proposes 'to ensure that the history of cricket and the spirit of the game, its rich heritage and culture, are built upon in a contemporary manner'. Here one could do with some further explanation – how do you develop a regard for tradition and custom 'in a contemporary manner'? By penalising those who deviate from it? That, surely, is policing statute not inculcating spirit. By quarantining players from contact with potential corrupters behind state-of-the-art security cordons? This supposes a spirit so fragile it's scarcely worth preserving. By promoting the use of an umpiring panel and electronic aids in officiation? That's 'contemporary' all right, but out of keeping with history, spirit, heritage, culture.

While we're at it, whose history and whose spirit are we discussing here? That, it seems, is the ICC's business. Maintenance of cricket's values was once a collective enterprise – a trust devolved to each individual cricketer. Remember the famous injunction of Lord Harris, which Sir Donald Bradman was wont to invoke: 'Foster it, my brothers, so that it may attract all who can find time to play it; protect it from anything that would sully it, so that it may grow in favour with all men.' Well, bollocks to that: the ICC has now arrogated all fostering and anti-sullification responsibilities to itself, with what seems an apathetic shrug from the rest of cricket. The two, in fact, have been curiously interwoven: the pages on its website devoted to 'The Spirit of Cricket' describe the council's development program in countries new to the game, with the unhappy implication that the spirit is defunct everywhere else.

For some, especially those who once disdained the ICC as a tame cat and deplored the ambiguity of its status, this should be a cause of celebration. At least someone is taking charge. Yet ambiguity has been replaced by … ambiguity. There's more of what George Bush Snr once called 'the vision thing' in 'ICC Vision 2005', which is the body's stated set of objectives for two years hence. Actually, they're not so much objectives as subjectives, and if not contradictory, replete with tensions. The ICC intends to be 'authoritative and willing to make hard decisions', 'clear, decisive and consistent', 'proactive and in control', yet also 'united and inclusive' and 'approachable and accessible'. War is peace in this most democratic of dictatorships.

Just as Orwellian, and perhaps even more elusive, are what might be classified as the ICC's prime directives. In its set of values, the ICC commits 'to focus on the common goal of achieving the best for world cricket regardless of individual aspiration' (we've got our eye on you, Mr Dalmiya). ICC Vision 2005 translates this into the goal: 'At all times to act in the best interests of world cricket.' To which the only possible response is: what are they? Again, in the absence of further and better particulars, the answer seems to be: whatever the ICC decides.

Mission statements, lists of values and visions are intended to clarify. These obscure. Their only outcome is to preserve the ICC's discretion to act when it deems the 'best interests' of cricket at stake: the ICC proposes and the ICC disposes. This is confusing. As previously observed, the best interests of a game are not always easy to discern, and those choices that enrich the exchequer seldom come without cost. And, sadly, when an authority controls the making of decisions about both commerce and convention, the modern trend is for the latter to run a long way second: measurable gain always wins out over impalpable loss. Mr Speed, who as CEO of the ACB oversaw the discarding of the Sheffield Shield after a competition had begun for which it was the stake, might be thought to have a track record in this respect.

Rather than a past example, however, consider a live one – one that in the next decade, probably around the time of the next World Cup, we are bound to broach. Let us say it is decided that international cricket must tackle the United States. Let us say it is proposed that cricket's laws undergo some changes in order that it might appeal to American sports consumers. Let us say they are profound. What are cricket's 'best interests' here? The US, some believe, is cricket's candy mountain; from its summit, many hitherto inaccessible places would be visible: Mr Speed remarked at the ICC's annual meeting in June that cricket's becoming 'a successful niche sport' in the US 'would be of great benefit to the game globally'. But at what point would the conditions of cricket's sale trammel the property itself? The ICC's creed offers no guide, nor even the hint of a guide. Nor does cricket's capacity for insight into itself offer much hope.

The gerontocracy who once ran cricket had their faults to be sure: they were scarcely inclusive, far from accessible, and a million miles from approachable. But what do they know of sports marketing who only sports marketing know? Keeping bad ideas out of cricket, to appropriate a line of *New Yorker*'s baseball sage Roger Angell, is like defending democracies in South America: 'Juntas and re-schedulers are always hovering in the shadows, and when at last their shiny new regime is given a trial term in office one somehow knows it is forever.'

This should not be interpreted as another *Wisden* jeer for the Speed machine. Our English counterpart has in any case already given its views ample vent, which Mr Speed paid back with interest by dismissing the almanack as 'irrelevant'. Withal one of the benefits of irrelevance is the scope to speak freely. These are questions, in any case, important for all involved in cricket government. Why are they there? What outcomes are they seeking? How will they decide when they have succeeded? If they don't know that, they'll have their work cut out averting failure.

Wisden Australia 2003–04

Australian Crowds
Noises Off

'GO HOME LOSER. DON'T COME BACK.' When New Zealand's Roger Twose arrived in Hobart as a replacement on a tour of Australia a few years ago, he picked up his luggage at the airport to find this scrawl on his cricket coffin, apparently the judgement of an under-employed baggage handler. Now you know why those carousels take so long: your bags may be awaiting personal endorsement.

You may also have a glimmer of the nature of touring Australia. Australian teams have been among the first to bemoan the hostility of conditions elsewhere; it's seldom remarked how harsh is the initiation awaiting their guests. They used to send convicts here with ball and chain; it's not much easier for cricketers sent with ball and bat.

The runs are on the board. Never mind Australia's recent record for the moment. The fact is that in 126 years of Test cricket, Australia has lost only five home series to nations other than England: four to the West Indians in their pomp, one to New Zealand by the odd Test. Not all that is explained by sheer cricket excellence. The fact is that Aussies walk tall in their own land; others tend to take at least a haircut.

Australia would be a cricket challenge were the arenas empty and locals sensitive, multi-lingual polymaths. Across this huge island continent, visitors encounter every conceivable environment: from Perth's hard tarmacs and harsh light to Sydney's dusty decks and enervating humidity. From a distance, the landscape seems undifferentiable; in fact, there are many Australias. What changes less are Australians, with their sense of superiority by association with sporting teams that are often the world's best. 'Your basic Aussie is a winner ... and second is a bag of horse manure,' New Zealand's Ian Smith has observed. 'How many teams who have come up against Australia in any sporting code have been anything other than the underdog?'

Australian crowds have always stirred strong reactions. After England's tour of 1897–98, defeated Drewy Stoddart deplored 'the evil of barracking' as 'no good to man or beast'. Percy Sherwell, by contrast, found the audiences endearing when he led South Africa round Australia in 1910–11; 'to play without the accompaniment of cheers and groans and advice gratuitous,' he opined, 'would be to eat an egg without salt'.

That Stoddart found more to complain of than Sherwell makes sense, for Australians have tended to save their harshest receptions for their oldest rivals. English leg-spinner Cec Parkin found fielding before a packed Australian house eighty years ago an experience like no other: 'A crowd of fifty thousand sits in the terrific sunshine. To see a Test match, Australians have been known to travel a thousand miles. Work is suspended in the afternoons, and a contrast with the packed ground is the quietness of everywhere just outside … If you make a mistake, you have to go through it. I remember during the First Test match I somehow got fielding in the long-field. The crowd just behind me kept shouting: "What's your name, cocky? Who said you could play cricket? It's a rumour."'

By the time the first post-war Englishmen arrived, it felt like the whole country was part of a campaign of demoralisation. Their skipper Wally Hammond recalled the contempt even of the children who gathered round the team's first net session in Perth: '"That Hammond's not half the size they said he was!" growled one disgusted youth … "This Edrich don't look much – pretty weak off the back foot," declared another child of about four feet nothing … And another: "I don't like the look of Yardley's wrists – look weak to me" … Best of all was the eight-year-old … who critically watched Wright bowling, clicked his tongue in despair, and said: "I'd rap the pickets with that myself Pommy; you'll never move Bradman with that stuff."' Australians can be jarring even in ways they don't mean. Mike Brearley once described how gruelling receptions could be when one was on the receiving end of a string of bear-like handshakes. He unwound after the 1978–79 Ashes series with ten days in India, 'where the handshakes, like the culture, are more gentle'.

Not that Australian audiences were entirely uncharitable. They toasted plucky underdogs, like the 1950–51 Englishmen and the 1960–61 West Indians. After leading India here in 1967–68, 'Tiger' Pataudi wrote fondly of 'those friendly and hospitable Australian people who work and play hard, but know how to relax'. But as Richard Cashman observes in his *Ave a Go Yer Mug: Australian Cricket Crowds from Larrikin to Ocker*, 'partisan behaviour became the norm in the 1970s'. Australian nationalism was finding in general a more raucous voice, and it was exercised in no context more freely than sport.

An illuminating passage in Colin Cowdrey's *MCC* evokes the chauvinism of Australia's 1974–75 series against England – the high summer of Dennis Lillee and Jeff Thomson. Australian golfer Peter Thomson dropped in to England's dressing room one evening during the Melbourne Test as the visitors played out another day under fire from both fast bowlers and fans, and shared with Cowdrey his own recent experience of crowd partisanship during a play-off against an American opponent in the US. Thomson explained that he'd hit a seven-iron into a bunker, and been astonished to hear a huge roar of delight. It had been, he told Cowdrey, an epiphany: 'What I had to realise was that crowd values have changed. They are not always going to lean over backwards to show generosity to the visitor … I had to take a grip on myself and realise that we want those huge crowds lining the fairway, we want that kind of enthusiasm, we even want that kind of enormous spectator participation. The crowds pay their money and they are demanding their pound of flesh. It is something that cricketers, like golfers, will have to learn to live with.' Sportsmen of all stripes since have been unconsciously echoing Thomson's stoical sentiments.

Australians scarcely have a monopoly on boorishness. To spectators the world over, the freedom to scorn and ridicule seems to be considered an inalienable right: even Scotland at the 1999 World Cup had their 'Tartan Army'. Yet perhaps because they come together in such numbers and with such expectations, Australian crowds can be especially menacing. Recalling his Test debut at the MCG thirteen years ago, Phil Tufnell described how the magnitude of an English batting calamity was amplified by the accompanying ululations: 'I have never been so frightened at a cricket match before or since, nor more disappointed. The feeling that this was all slipping out of our control, and the reaction of the crowd, increasing by a notch on the dial with every dismissal, combined to unleash an almost tangible sense of fear.' Each year there seems some slightly uglier development. When Muttiah Muralitharan was no-balled at the MCG on Boxing Day 1995, the shock that rippled round the arena was full of fellow feeling; since then he has been baited, sickeningly, with cries of 'no ball' at almost every appearance.

One-day games often get the blame for crowd degeneracy. This is not quite fair. In some senses, limited-overs cricket provides a place for bone-headed spectators to congregate, leaving the five-day form to proceed in relative calm: recent Tests at Melbourne have even seen some niceties restored, like applause when bowlers end their spells.

Nonetheless, one-day fixtures seem to have less and less to do with cricket, and more to do with crowds making their own 'entertainment'. It was a shock for Australians, but no-one else, to learn last December that the International Cricket Council regarded the MCG as one of the world's worst three venues for interruptions in play, trespassing and unruly behaviour; season 2001–02 saw thirty arrests and 500 ejections. Cricket Australia chief executive James Sutherland made the shamed admission: 'There is no doubt that the ICC have taken a much stronger view of poor crowd behaviour and they will shortly be introducing the powers to ban grounds where they don't meet the standards or they have a record that has been inferior to the required standards in the past.' Shane Warne, as ever, amiably missed the point: 'I think the Victorian crowd and the MCG crowd are absolutely sensational. I think they get right behind Australia.'

What to do? Siege mentalities develop quickly. Touring teams have been coming to Australia for years now and communicating their unease about playing here; Steve Waugh's team have been experts at exploiting their discomfort. Waugh himself exemplifies the opposite approach. A taciturn, introverted young player, his increasing cos- mopolitanism as an elder statesman was recognised in September by his appointment as an official Australian tourist ambassador in India.

'Teams get into trouble overseas when all they do is sit around in their hotel rooms,' Waugh believes. 'You've got to enjoy touring or it's going to become a chore ... I really think that's been the secret of my success away from Australia, that I've learned to go out and enjoy the places I've been, which means that you're not so focused on how homesick you are or what your form is like.' If you don't try to enjoy touring Australia, you won't enjoy it. And if you come expecting to hate it, you will really hate it.

Wisden Asia October 2003

Umpiring
Inglorious Uncertainty

'ACCURATE UMPIRING AT LAST!' So read the headline on the cover of English magazine *Wisden Cricket Monthly* six years ago, acclaiming the adoption of video arbitration as a landmark in the game's history. We're still waiting. The last five weeks have been a setback for the umpiring profession: farcical run-outs, disputed catches, botched boundaries and a bowler with perhaps the most scrutinised action in history in Sri Lankan Muttiah Muralitharan whose technical purity remains nonetheless indeterminate.

This summer's tourists patently don't feel they're being better umpired than six years ago. The visiting Englishmen were aggrieved by decisions given against them by third umpires Paul Angley and Simon Taufel during the Ashes series, while the visiting Sri Lankans appeared on the point of lynching umpire Ross Emerson last Sunday when he had the nerve to no-ball Muralitharan for throwing. One recalls the dictum of American scientist Paul Ehrlich: 'To err is human, but to really foul things up requires a machine.' So why is the modern cricket umpire at the top level – with his walkie-talkie, video aids, National Grid sponsorship and International Cricket Council referee – still apparently arbitrating with such inglorious uncertainty?

Time was, of course, when the man in white was always right. Writer Jack Fingleton whimsically observed forty years ago that Test umpires were even more powerful public figures than politicians, able with a single gesture to send nations into hysterics of grief or rapture. Such authority, however, was never destined to last. As Pakistan's former captain Asif Iqbal once asked: 'I don't understand why, in a democratic society, where government and all the accepted standards in every walk of life are being questioned, why umpires should be immune.'

As times have changed, so have views. In his first book in 1980, for instance, English all-rounder Ian Botham stressed: 'Never argue with an umpire. There will be times when you want to ... but the laws of cricket say the umpire's decision is final ... If you stand and complain you only make yourself look silly.'

Yet who should be umpire Emerson's most astringent critic a week ago and the Sri Lankan captain Arjuna Ranatunga's most outspoken supporter during their Sunday Punch and Judy show? Step forward, Ian Botham.

Did technology, in the form of replays that allowed us to second-guess every decision, hurt the umpire's cause? It could hardly have helped, especially when flashed up on giant screens for the crowd's edification. Some umpires have professed to feeling unthreatened by replays. But even Test cricket's most experienced umpire, Dickie Bird, has confessed that he did not in a twenty-eight-year career ever glimpse footage of one of his arbitrations: 'I never watch myself umpiring whether on the edited highlights at night or on the Nine O'clock News. If I did, I would probably have gone crackers by now.' A Dickie Bird even more crackers scarcely bears thinking about.

Technology, however, also held out the promise of re-establishing the umpire's stature. With TV aid, it was said, the umpire's grip would be strengthened: no longer would their errors be ridiculed; no longer could players feel disgruntled.

It hasn't worked out this way, and this summer has been no dif-ferent. In the Adelaide Test, England's Michael Atherton edged low to slip where Mark Taylor claimed the catch. Third umpire Angley, who had only umpired three first-class games, upheld the appeal in a blink. While consenting that Atherton was probably out, *Wisden Cricketers' Almanack* editor Matthew Engel made swingeing criticism of the swiftness of the decision in his Test report. 'I reckon it was blind panic,' he concluded.

In the Sydney Test, Michael Slater was apparently found short of his ground with his score on 35. Umpire Steve Dunne referred it to third umpire Simon Taufel but, as amply reported, the cameras were not perpendicular to the crease lines and the batsman enjoyed the benefit of the doubt. Slater went on to 123, the decisive innings in a low-scoring match. The two decisions, while as different as investigations by Inspector Clouseau and Inspector Gadget, contain a common thread: they suggest that subjectivity in cricket adjudication is still present, even when replays are involved, because the officials involved apply different burdens of proof.

Another question to arise from the Slater run-out involved the referor rather than the referee. Consider this description of the method of judging a run-out at the non-striker's end by English umpire Don Oslear:

> Firstly, you have to be a world-class sprinter over ten to fifteen yards. Then you have to be in line with the batting crease, and standing still. At times, you have to be something of a gymnast. My personal *modus operandi* … is not to take my eye off the ball until it has left the fielder's hands; then if necessary I spin around, sometimes while airborne, and focus my eyes on the line.
>
> When in position … I then go down on one knee. There are two reasons for this: (1) if the throw is from behind me it will be able to pass safely over me, (2) I can better observe if the bat is in contact with the ground behind the batting crease when the wicket is broken.

Adjudicating Slater's run-out, however, Dunne worked himself into a position from which he had no chance of making a decision: standing erect to the off-side of the pitch for a throw from long-on so that the wicket was invisible to him, and perpendicular not to the crease but to the stumps. And this on a *second* run.

The question crossed at least a few minds in Sydney: is the third umpire's availability having deleterious effects on umpiring technique? It wouldn't be the first occasion on which the introduction of labour-saving technology has caused human skill to atrophy.

The case of Muralitharan, meanwhile, is murkier still. It might be instructive to recall the ways cricket has hitherto dealt with bowlers of suspicious actions. The first concerted *fatwa* against throwers, in England at the turn of the century, was initiated by a coalition of strong-willed umpires and county captains. The second, internationally in the late 1950s and 1960s, sprang from a common objective fostered again by umpires and senior administrators of Test-playing nations.

Technology played a role in the latter case, England's top administrator Gubby Allen commissioning photographer Ken Kelly to film a number of suspect bowlers secretly with a 16mm cine-camera. But no-one disputed Sir Donald Bradman's sage words on the subject: 'The throwing question is so complex because it is not a question of fact but a question of opinion or interpretation. Men of unquestioned goodwill and sincerity take completely opposite views. The laws say the umpire should judge the issue – if anyone has a better suggestion, I'd like to hear it.'

No-one, apparently, is satisfied with Sir Donald's trust in umpires today. And who needs 'men of unquestioned goodwill and sincerity' when you have five lawyers to keep Arjuna Ranatunga from harm? Yet the fact that we are no closer to a precise judgement of whether Muralitharan obtains an unfair advantage by his action for all the video-aided scrutiny of commentators, committees, doctors and bio-mechanists implies that the first part of his statement holds true: we may be looking for an indisputable scientific 'fact' that isn't there.

Not all has been gloom for umpires this season. In between times, there has been a Melbourne Test adjudicated quite wonderfully in the trying circumstances of a tense and draining game by West Indian Steve Bucknor and Australian Daryl Harper. Of course, they were scarcely mentioned in despatches, for Sir Neville Cardus's remark about umpires still holds true: 'The umpire at cricket is like the geyser in the bathroom; we cannot do without it, yet we notice it only when it is out of order.'

Nor can one conceivably turn back the clock to before the age of video and rely again solely on the naked eye. Applied with care by well-trained and experienced officials, the third umpire stands to benefit cricket considerably. It is already having an effect on the techniques of the game. In a recent interview with *South African Cricket Action*, the Proteas' in-field assassin Jonty Rhodes commented that he now feels able to slow down and take more time over aiming for direct hit run-outs, knowing that where once he had to catch the batsman at least a foot short, now it need only be a centimetre.

Yet there is a lesson to the umpiring schemozzle of this summer, and that is that subjectivity in adjudication is ineradicable. This we once condoned, trusting sufficiently in the *bona fides* of umpires to treat the man in white as right even when wrong. It didn't always work as a principle, and was never likely to survive the pressures of modernity. But once you discard such a convenient and sustaining understanding, with what do you replace it?

The Age January 1999

Throwing
The Great Taboo

CRICKET'S OCCASIONAL ALARMS ABOUT illegal actions, disruptive as they can be, also attest one of its great successes. Some bowlers have, at various times, transgressed the law; not merely most but almost all have remained well within it.

This is not a minor consolation. A vestige of when bowling was 'bowling', delivered underarm, projecting the ball with a rigid arm is not an action at which one would naturally arrive. It is far from clear how, when and why it became convention; even Rowland Bowen in his mighty work, *Cricket: A History*, can do no better than: 'At some unknown stage, the idea took root that "cricket" bowling involved a straight arm.' And yet we do it, with such fastidious fidelity that anything other seems an offence to truth and beauty.

This is all the more surprising when it is considered how easily a bowler might transgress inadvertently, as the Australian umpire Jim Phillips pointed out just over a century ago: 'Just as one bowler, in his desire to make his delivery more difficult, gets as near the return crease as possible, and occasionally inadvertently oversteps the mark, thereby bowling a "no-ball", so another bowler will, in attempting an increase of pace, use his elbow, especially if he be a bowler whose arm is not quite straight. In each of these instances it does not seem just to suppose that either bowler is wilfully unfair.' And Phillips might be thought to know whereof he spoke, being at the forefront of the first of cricket's three most serious and protracted problems with illegal actions (I am here excluding the earlier complaints of 'throwing' associated with the trial and adoption of roundarm and overarm bowling). The common qualities of each crisis are instructive – the differences even more so.

For some time after the legalisation of overarm bowling 140 years ago, very little was done about those at the edge of orthodoxy – which is unsurprising, given that so much previously dubious was now *de rigueur*. Lancashire seemed positively to breed them. John Crossland was finally exiled from the game not because of the protests about his action from Lord Harris but because he was discovered to have breached his residential qualification, living in Nottingham for a period in 1884; his contemporary George Nash, a left-arm spinner whose wickets cost only 12.3 apiece, and successor Arthur Mold, who took 100 wickets in a season on nine occasions and was probably the fastest bowler in England, shied with impunity for many years.

By the early 1890s, throwing was disturbingly pervasive in English cricket, and being assimilated by impressionable colonials. Erstwhile Australian skipper Billy Murdoch heaved an exasperated sigh in *Wisden*: 'To my mind the remedy for throwing is very simple; the rule says "if the umpire be not satisfied of the absolute fairness of the delivery of any ball he shall call 'no ball'." Now if the MCC were to instruct all umpires that it was absolutely necessary to carry out this rule without respect to persons, I think it would have the desired effect.' But in the absence of firm policy, actions continued to mutate. As Fred Spofforth put it in a celebrated letter to *Sporting Life* in January 1897: 'There is scarcely a first-class county which does not include a "thrower" amongst its cricketers, many of them men who would scorn to cheat an opponent out, and who, if a wicketkeeper were in the habit of kicking down the stumps or knocking off the bails with his hands … would not hesitate to bring him before his committee … Still they will not only employ a man to throw, but will actually throw themselves, and acknowledge it, their only excuse being that "others do it"…' Does this sound familiar?

Then came Phillips. A peripatetic character born in dusty Dimboola who trained and practised as a mining engineer before and after his cricket days, he was a unique freelance official who commuted between Australia and England in order to ply his trade long before it became fashionable: he was thus unusually well qualified to see that deteriorating bowling actions were not a problem exclusive to either country, and exceptionally well placed to do something about it. After Phillips' first proscription – Australian Ernie Jones at Adelaide in January 1898 – *Wisden*'s Sydney Pardon predicted: 'From what Phillips has done, nothing but good can come.' He was right. During the three English seasons from 1898, Phillips and emboldened contemporaries passed sentence on half a dozen serial offenders – repeatedly where necessary, as it was in the case of Mold. If it did not disappear, the problem faded.

Throwing's recrudescence fifty years later was in an environment somewhat more complicated. There was again a squeamishness about responding in England, but this time no umpire came forward to take the initiative: Frank Chester was of a mind to call the South African Cuan McCarthy at Trent Bridge in June 1951 but refrained when the aqueous Sir Pelham Warner indicated that he would receive no support from the MCC. And thanks to similarly invertebrate administration elsewhere, the contagion spread. By the late 1950s, virtually every country had a group of bowlers with actions of arguable purity, including some of the fastest of their era: West Indian Charlie Griffith, South African Geoff Griffin, New Zealander Gary Bartlett, and the Australians Ian Meckiff and Gordon Rorke. 'The danger of not stamping on offenders in the past has led to the problems which now confront the authorities,' wrote *Wisden*'s Norman Preston. 'They have only themselves to blame ...'

Remedial policy required the building of a coalition of the willing, as it were, at the July 1960 convention of the Imperial Cricket Conference. Griffin had just been called eleven times in a Lord's Test, and Meckiff and Rorke seemed destined for the same fate the following season when Australia visited. In the event, for reasons of form and fitness, neither was selected. But the revisions made to Law 26 to prohibit 'sudden straightening' of the arm 'immediately prior to the instant of delivery', to curb dragging of the back foot, and to stand behind umpires choosing to enforce both statutes, provided at last the basis for coherent action. Sixteen different bowlers were called between 1960 and 1964, in England, Australia, the West Indies and Pakistan.

It is in comparison with the controversies fifty and a hundred years ago that the nature of the current impasse seems especially disturbing. Unilateral action by a strong umpire *à la* Phillips has been deemed unpalatable; multilateral action *à la* the Imperial Cricket Conference doesn't seem an option either. But the dispute can't be sidestepped. The chief *dramatis personae* are not consistent first-class performers like John Crossland and Arthur Mold, or peripheral Test players like Geoff Griffin and Ian Meckiff, but the world's leading Test wicket taker, Muttiah Muralitharan, and fastest bowler, Shoaib Akhtar. Both, to complicate matters still further, are unusual physical specimens anyway: Murali's elbow is congenitally deformed; Shoaib is elastically double-jointed. And biomechanic technology, to which cricket's governors turned in search of certainty, has simply introduced another layer of uncertainty. Precision in judgement on the question turns out to be asymptotic: it can be approached but never reached. Degrees of tolerance for arm straightening have been introduced in the light of studies revealing that most bowlers do, quite unconsciously and naturally, flex their arms while bowling. This, however, is little use to umpires. The choice seems to be between a law that is enforceable and unacceptable or one that is acceptable and unenforceable.

The overarching difficulty is that, after a hundred years in which authorities have reacted only with the greatest reluctance and queasiness, throwing has become the cricketing deed that dare not speak its name. The perspectives with which Jim Phillips enforced the law and with which some recent commentators have interpreted the law are quite different, and reflect, in some ways, a melancholy development. We live in topsy-turvy times. To bowl a string of malicious bouncers trying to pin a tailender backing towards square leg these days is to show commendable commitment and aggression; to bend your arm as a quick bowler by more than ten degrees or as a spinner by more than five degrees is to be a cur, a cheat and a blight on the game. There used to be a distinction in cricket between the legal and the ethical – but not, perhaps, anymore.

The Wisden Cricketer June 2004

The Allan Border Medal
Playing with Ourselves

DURING THE 1960s, Queensland's Peanut Marketing Board endowed an annual prize for the Australian Cricketer of the Year that, not surprisingly, was known colloquially as the Golden Peanut. Winners included Brian Booth, Neil Hawke and Doug Walters.

Who remembers it now? Awards are necessarily of the moment and, often as not, don't stand the test of time. Australian Rules football has its Brownlow, but it also once had its Cazaly, and nobody misses it much.

Monday night's inaugural Allan Border Medal had an air of permanence to it; the sort of swaggering confidence that comes, no doubt, from being lavished with the attentions of Channel Nine. Yet it also had other airs, less fragrant altogether. The structure creaks, and not merely from unfamiliarity. The vote-count rigmarole, so successful at the Brownlow from which it is obviously derived, is here no more than *faut de mieux*. Test cricket especially, so various in its conditions, so teeming with possibilities, doesn't seem well-suited to votes of the 3–2–1 kind. It seems odd, for instance, that Damien Fleming received the same number of votes for obtaining 3/14 in the rain-ruined match against Sri Lanka at SSC in Colombo at the start of October as Adam Gilchrist did for his imperishable innings against Pakistan at Bellerive Oval six week ago. And that Steve Waugh's breathtaking 120 in the last of the World Cup Super Six matches against South Africa profited him less, once discounted for occurring in a one-day match, than his twin's 90 at Harare in the inaugural Test walkover against Zimbabwe.

The Brownlow vote count, too, manages to attain a surprising degree of tension because the football season is essentially straightforward: a set number of uniform rounds. The Allan Border Medal features two separate strands of the game, judged independently, then reconciled arbitrarily, with media and umpires consulted as an afterthought for no particular reason. Tension? You wouldn't have heard a coffin drop, let alone a pin.

The concept of the players primarily judging themselves has merits and demerits. The jury-of-peers system is highly suitable to an award that bears the name of Allan Border: a cricketer's cricketer probably more esteemed by comrades than critics. Players' judgement might theoretically be clouded by rivalries and petty jealousies; but no more, and probably less, than the judgement of writers, officials or a panel of experts.

Yet there is a danger that, because electorate and candidates are the same, and a relatively small group at that, the evening itself mutates into merely the self-glorification of an elite. And there were times on Monday evening when proceedings resembled my old footy club's pie-night, with everyone sitting round over a lot of beers telling each other how good they were; a traditional and quite congenial Australian rite, but not something you'd necessarily televise. At worst, in fact, it manifested some rather unappetising aspects of Australian sporting success: tendencies to jingoism, self-regard and inwardness that we'd do well to resist.

Cricket needs the Allan Border Medal to succeed, to achieve the cachet and credibility that turn it into a fixed point in the helter-skelter cricket calendar. The cricket season has no obvious climax; summers more or less peter out, leaving it to the commencement of the various football seasons to signal their end. But an exercise in self-congratulation and big-noting for a cosseted clique would be worse than nothing, and deserve to die as unlamented as the Golden Peanut.

Wisden Online February 2000

Sledging

F*** It

I AGREE WITH STEVE WAUGH THAT the media is unduly obsessed with sledging in international cricket, and that less occurs than is popularly believed. But I don't accept that it's 'part and parcel of the game', as Australians are wont to put it, any more than I accept that road rage is part and parcel of driving. It occurs to me that Test players would sometimes benefit from seeing themselves as others see them. When Brett Lee celebrates bowling a tailender with a volley of abuse, I don't think: 'What an excitement machine!' I think: 'Plonker.'

Cricket's real sledging problem, however, is in the 99.9 per cent of the game that isn't international, isn't on television, isn't sponsored. I've been nine years at my present club in Melbourne. Each season the sledging has grown more pervasive, more personal, more pernicious, and more planned – to the extent that the possibility of physical violence has sometimes seemed far from remote. And here Steve Waugh does bear responsibility. The perception that what he euphemises as 'mental disintegration' plays a significant part in the modus operandi of the world's best team is highly influential on all behaviours.

Nobody expects cricket to be deodorised or bowdlerised. And frankly, you get used to sledging in Aussie club cricket, just as using email inures you to spam. It's gotten to the extent, in fact, where sledging might have had its day; it's now just a bore rather than a real competitive edge. Even in the humble grades I play, sledging seldom discernibly puts anyone off their game. It merely fosters an atmosphere of distrust and ill-feeling. Which is why I find its apologists absurd. Does it enhance anyone's enjoyment of or love for cricket? Does it contribute to mutual respect, understanding and fellowship? If not, while I can see an argument for tolerating its natural occurrence in a game played between adults with a hard ball under a hot sun, I see no grounds for defending it, and sound reasons for its extirpation.

I'd happily concede that cricket should always have a place for aggression: a game must test mental as well as physical prowess, lest it be merely an endurance contest. But I don't regard sledging as testing anything except our ability to dehumanise and desensitise ourselves and each other, nor do I see cricket as an arena in which aggression should simply go unchecked. A host of sports pander to our cravings for violence and excess already. Why would cricket want to be anything like them? In general, I find it strange that cricket should be apologising so consistently for not being like other games, that the idea of it having a spirit, and the notion that players should exercise self-control, should be thought of as archaic. I'd like, in essence, cricket to have the courage of its difference.

Wisden Cricket Monthly August 2002

Zimbabwe
A Test of Integrity

GEORGE BERNARD SHAW ONCE SAID that an Englishman thought he was moral when he was simply feeling uncomfortable. The same, increasingly, seems to apply to our sportsmen and sporting administrators. How else to explain how a team of Australian cricketers, queasily conscious of the dubiousness of the honour, were last week guests of Robert Mugabe?

Sport has hardly been declared morality free. We have furious, foam-flecked debates about matters that it would be flattery to call trivial: whether James Hird should have dissed an umpire, whether Sam Newman should be permitted in public without taking his medication. We nonetheless stumbled into Zimbabwe claiming heavy hearts and admitting utter confusion, while still mouthing the piety that 'it's only a game of cricket' – to quote Adam Gilchrist.

Gilchrist's baffled column in *The Age* last week is perhaps the most telling artefact of this sorry affair, having recourse in discussion of Zimbabwe's plight to phrases like 'alleged heartache' and 'reported suffering'; one awaits references to 'apparent bombings' in Iraq and 'rumoured detention' of asylum seekers. Zimbabwe's eleventh-hour cancellation of the Tests provides only a very small fig leaf to cover a very large embarrassment. The one-day internationals will be played as planned. Everyone will hold their nose to escape the stink of hypocrisy.

It should be clear that our cricketers found themselves in Harare not because they particularly wanted to be, or to further cricket's good name, or even because they were expecting a good game – quite the contrary. Because the internal exile of the country's (mainly white) first XI has left a ragged (mainly black) second XI in its place, a catch-weight contest had been expected.

They were there, instead, because of the consequences of their not going. Swingeing fines and a costly suspension await those member countries of the International Cricket Council breaching contractual undertakings to tour others; time will tell how it deals with those who rescind invitations. Cricket Australia described the exercise as 'a box we have to tick'. Boxes are an important cricket accessory, but not usually in this respect. It was a dismal auspice for a cricket tour.

This being Australia, from where the rest of the world is viewed as through the wrong end of a telescope, there has been precious little information about Zimbabwe's benightedness as a nation. Free speech and fair elections are things of the past there; dissent is ruthlessly crushed by means from expropriation to execution. By every conventional measure, the country is sliding backwards. Life expectancy is lower than in 1960. Two-thirds of its population is on the brink of famine or 'food insecure', despite three-quarters of the country's grain already coming from the World Food Program. About three million of its people have fled, mostly to South Africa. The economy has contracted for four consecutive years. Inflation runs into hundreds of per cent; unemployment, officially fifty per cent, is probably higher. Nor are these acts of God; they are acts of Mugabe's corrupt, despotic and increasingly desperate regime in no more than a handful of years.

It has been reasonably pointed out that we play sport against countries that aren't exactly democratic fashion plates. What made this tour farcical, however, was the way Mugabe's ZANU-PF has so unceremoniously wrested control of the Zimbabwean Cricket Union in the last few years, populating it with cronies, politicising its management and selection processes, and turning it into a vehicle for propaganda and personal aggrandisement.

Zimbabwean cricket's travails have hardly been secret. Since 2000, twenty leading players have left Zimbabwe prematurely including Neil Johnson, Murray Goodwin, Andy Flower and Henry Olonga. The rest, courageously, held their peace until their captain Heath Streak made a private protest about various recent selections that on

2 April earned him the sack; they were then sacked for expressing their disapproval.

This was depicted to the world as a matter of race, which has a way of making liberal consciences quail. But what ZANU-PF has wrought in Zimbabwean cricket has been only at the margin about the empowerment of black countrymen, or toppling a final graven idol of colonialism. As with so much of the Mugabe 'revolution', it has been about party apparatchiks seeking stature, and another exchequer to raid, amid encircling chaos. Remarkably, twelve ZCU members managed to visit Australia during Zimbabwe's tour last season, with partners, all expenses defrayed – remarkably because the ZCU struggles to keep a viable first-class cricket structure going in Zimbabwe. We should not only never have set foot in Zimbabwe, we should never have come close.

Such debate as there was about the advisability of Australia's tour, alas, was crushingly disappointing. Apologists ran the tired line that politics and sport should not mix; like it or not, they do, time and again, and it is the ZCU that has in this instance done the mixing. A counterargument was that cheap runs and easy wickets on offer endangered the sanctity of cricket statistics, as though recalcitrant scorers might ignite an *auto-da-fé* fuelled by *Wisden*s.

It is strange to recall that cricket was once supersaturated with morality – sometimes to the point of nausea. These days, professional athletes are not expected to think about anything more than their sport; everything else is done for them, even their autobiographies. 'The bottom line for me is that we are cricketers. Our job is to play cricket,' wrote Ricky Ponting in his most recent book – or at least so said a ghost writer on Ponting's behalf.

Curiously, when it suits them, athletes love cloaking themselves in the flag, proclaiming the patriotic pride they derive from representing their countries – and *we* love it when they do. Hundreds of millions of Indians and Pakistanis who revelled in their countries' recent Test series, for instance, did not think they were watching twenty-two men doing their jobs. If Zimbabwean cricket was merely a place of work, ZANU-PF would never have coveted its control. Professional athletes should be careful about drawing too much attention to that 'bottom line', lest they be taken at their word.

For here lies, for sport, perhaps this fiasco's most troubling dimension: not merely that moral arguments no longer have traction, but that they seem to have been supplanted by commercial considerations. Over the last seven weeks, the International Cricket Council has kept an incriminating silence, save for one ill-aimed volley from president Ehsan Mani on 7 May: 'If the rebels believe that walking out will result in other countries interfering in Zimbabwean cricket, I think that they have been very badly advised.' A strange world this when players sacked and expelled become 'rebels' who are 'walking out'.

Then on 18 May, after the ZCU had refused to meet his chief executive, Mani foreshadowed a meeting of the ICC executive board that threatened the tour's official status: 'It's up to the directors to determine if these matches should have Test status or not and to exercise their judgement as to what course of action best protects the integrity of the international game.' V.V.S. Laxman could not have executed a glide more effortless. One moment the ICC could not be 'interfering in Zimbabwean cricket'; the next it had to protect 'the integrity of the international game'. For 'integrity', though, read: the value of the franchise.

In June 2000, the ICC signed a seven-year, US$550 million deal for the broadcasting of international cricket with Global Cricket Corporation, now an arm of News Corporation. But Test matches in which teams declare at 3 for 713 – as Sri Lanka recently did against the new Zimbabwe – and one-day matches in which teams are routed for 35 – as the new Zimbabwe recently were against Sri Lanka – scarcely enrich cricket's commercial cachet. It's unlikely Rupert Murdoch himself has been on the phone; more probably – like so many editors – the ICC has simply uncannily anticipated his concerns. This is what happens when it becomes too vexing to distinguish right from wrong, being moral from feeling uncomfortable: money makes the decision for you.

The Sunday Age June 2004

The Writing Game

E.W. Swanton
Lord Jim

ONE OF THE BETTER STORIES of E.W. Swanton, grandee of the English press box who died at the weekend at the age of ninety-two, concerns the Lord's Test between England and the West Indies in 1963, which coincided with a conclave of cardinals at the Vatican to select Pope John XXIII's successor. As the world awaited the traditional puff of smoke from St Peter's to signal the end of deliberations, BBC cameras homed in on a chimney that had caught fire on the old Lord's Tavern. The incorrigible Brian Johnston added the perfect caption: 'Ah, I see that Jim Swanton has been elected Pope.'

It's a typical fragment of Johnstonia, but also evokes something of Swanton's stature in the game even thirty-seven years ago. Nor was it merely through longevity that Swanton achieved such exalted standing; it was a combination of bearing, balance, soundness of judgement and cognisance of history that lent his words the authority of papal encyclicals.

John Ward's painting on the cover of one Swanton anthology, *As I Said at the Time*, shows the writer seated against the backdrop of a tree-lined county ground, notebook in hand. The attitude of ease, the setting, even the prop, could scarcely be bettered: Swanton's journalistic career began in 1927 with *The Evening Standard*, before the spread of radio and television, and his match reports retained always a clarity and comprehensiveness that make them a pleasure to read decades later.

As a prose stylist, in fact, Swanton makes a fascinating study. There have been many more felicitous phrasemakers and more memorable aphorists. Yet a passage of Swanton is genuinely unmistakable. To choose one Australian Test match report at random: 'It is one of the fascinations of cricket that it is so clear and open a test of character, and no-one wants to see the natural humours of a man subdued to the point of dull anonymity; but the truth is that these tremendous sporting affairs, on which it can be thought the prestige of nations rested, can only be kept in any remote degree of perspective if all concerned, players, press and public, do our best to see the other fellow's point of view, and, frankly, mind our manners.' It may also enhance the readers' appreciation of Swanton's vista of experience to know that these lines – carefully qualified but subtly admonitory – come from a report in Melbourne's *Argus* on the Third Test of the 1946–47 Ashes series.

Cricket will not see Swanton's like again. The resonant judgement has become the preserve of the big name ex-player. Cricket is also no longer so secure in its sense of shared values. Yet the game is the poorer for the loss of one so revered, and also so generous. Among the host of tributes collected and sifted by Cricinfo, there are comments about the encouragement he offered the young. His journalistic *protégé* in the early 1950s, John Woodcock, became one of the finest writers in the game and a splendid editor of *Wisden*. His personal assistant in the early 1960s, Daphne Surfleet, married Richie Benaud and became his consulting partner.

This is something, at an altogether humbler level, I can attest. A couple of years ago, after I published a book called *The Summer Game*, a friend sent a copy to EWS at his home in Sandwich with a letter noting that I had a hefty collection of Swantoniana. Not long after, a large envelope arrived in the mail with a friendly missive and a fistful of bookplates inscribed in a steady hand: 'Warmest congratulations, E.W. Swanton.' Feeling like a stage-struck extra who'd garnered a favourable notice from Alexander Woolcott, I pasted them into the appropriate volumes. It's pleasant to think of him commentating in heaven, perhaps alongside contemporaries like John Arlott and Alan McGilvray, with Brian Johnston keeping a watchful eye out for smoking chimneys.

Wisden Online January 2000

Roland Perry
No Ball

FIRST, A DISCLAIMER. Four years ago, I added some chapters to *On Top Down Under*, Ray Robinson's 1975 masterwork on the cricket captains of Australia, in order to bring it up to date and introduce it to a new audience. It was like returning, in small part, a great debt to a lifelong benefactor: no more assiduous and scrupulous work of Australian cricket history exists. I wondered accordingly, on learning of Roland Perry's plans to publish a book with the same subject, how he could possibly improve on Robinson's *magnum opus*. The answer is: he can't.

Perry emerged on the cricket-writing scene five years ago with *The Don*, a once-over-lightly Bradman biography, under-researched and overwritten. He begins *Captain Australia* in similar vein with a string of lazy mistakes, including in the first few pages that Henry Scott was the son of a Royal Marine (he was the grandson), that Warwick Armstrong was the son of a liquor merchant (he was the son of a lawyer), that Douglas Jardine was 'a high-born Scottish lawyer' (Jardine and his father were both born in India, neither were nobs), and that Australia might have drawn the 1938 Oval Test had Bradman been fit (the Test was timeless). All this alongside constructions such as 'Bill Woodfull, the non-drinking headmaster son of a preacher'. Ugh.

Thereafter, the error quotient falls, if only because the book is so bland and pedestrian. But Perry retains a disquieting tendency to, quite casually, mangle information for no particular reason. Sometimes, this is mere embroidery, though it is regular enough to be annoying. Sometimes, it is deliberate manufacturing, such as a description of William Murdoch's famous 153 at the Oval in 1880 that

Perry ascribes to Sir Home Gordon 'in a London paper': 'Murdoch was always graceful. He cuts beautifully ... He is a jolly genial man with much appreciation of other players.'

The quote, lifted from *On Top*, has been subtly doctored. Gordon was not describing Murdoch's landmark innings at all – being eight years old in 1880 – but writing generally some years after Murdoch's retirement. And what he actually said was: 'Murdoch was always graceful. He could cut beautifully ... He was a jolly, genial man, with much appreciation of other players and too little control of his own inclinations.' It appears that Perry has taken a retrospective judgement, changed the tenses and presented it as contemporaneous reportage. I guess it saved time.

Elsewhere, there are assertions whose origins are, at least, somewhat elusive. In the chapter on Joe Darling, for example, Perry tells us that he instructed players to wear their new 'velvet skull caps' throughout the Sydney Test of December 1901 'to present the opposition with the perception of a tight unit, truly representative of a new nation'. An interesting story. But no source is cited – it's not in any standard reference work, or in Darling's memoirs – and pictures of Australia in the field during the game show almost all of them wearing sun hats. Rather disobedient of them.

Likewise, there's the old chestnut about who leaked the dressing-room confrontation between Bill Woodfull and Pelham Warner during the Bodyline series. Jack Fingleton gets the blame again, *via* Australia's twelfth man Leo O'Brien. Says Perry: 'Seconds later O'Brien told Australian opening bat Jack Fingleton ("word for word" according to O'Brien) what had been said.' Again there is no source, however, while Perry ignores not only Fingleton's lifelong denials of involvement, but also the quite different version of events that O'Brien gave *Wisden Cricket Monthly* in April 1983 (where he mentioned that Fingleton was not in the room, but that several former players were, any one of whom could have spread the story). Perhaps the author knows something nobody else does; if so, it would have been nice to know on whose authority. Perhaps O'Brien told a different version of the story elsewhere; if so, he's scarcely a credible witness.

I could go on, but these remarks are merely by the way, for the main reason this book is so irksome is that it doesn't amount to a hill of beans. Where Ray Robinson sketched Australia's captains in the round, bringing them alive with vivid and intimate detail, Perry depicts them in the flat, bulking the thin personal matter with ponderous clichés, execrable puns, and recitations of scores and averages. There's no original research to speak of, no new insight to relate; there's not even a bibliography. Matthew Engel noted in *WCM* two years ago that 'cricket histories get written, well received, widely read, and sometimes even win awards, when the writer has done damn all except flick through a few old books'. I couldn't put it better.

Perry's biography of Steve Waugh, *Waugh's Way*, is not nearly so offensive. It's merely abysmally dull, which we're used to in Australia, where publishers seem to lower the bar annually on the standard of cricket books. Composed from secondary sources, it recapitulates games with sentences as gripping as: 'Waugh was economic, but not penetrating, taking none for 34 off 18 overs with four maidens.' As for Waugh's character, the judgements are as profound as: 'He despised "weakies".'

Ken Piesse's *The Complete Shane Warne* comes as something of a relief, in that it is a book that doesn't pretend to be anything it's not. It simply proceeds through Warne's bowling career match-by-match, the comprehensive statistics punctuated by brief narratives and appreciative quotes. It's not really history or literature, but it's fun, in the spirit of B.J. Wakley's *Bradman the Great*, and the records section is compendious. Warne's admirers would like to believe that a revised edition will be necessary within a few years. Time will tell. Sometimes you can improve on the already excellent. Sometimes, as Roland Perry demonstrates, you cannot.

Wisden Cricket Monthly January 2001

Tim Lane
Voice of Summer

IT IS A CHARACTERISTIC OF the cricket season that it never seems to begin. There is nothing to announce it – no opening ceremony, nor even a universal round one. International, domestic and club cricket all start independently. Not even the weather is a guide: summer sometimes cooperates; sometimes not.

The cricket lover needs other rituals and sensations of commencement. For my part, I never believe the season to be underway until I've heard Tim Lane's voice on the radio. As Alan McGilvray was to an earlier generation, so Lane has been to mine: the national witness, the acme of reliability.

I'll have to find another stimulus this season, though, because there is no Tim Lane. When he left the Australian Broadcasting Corporation earlier this year to call Australian Rules football for Channel Ten, he had to surrender his cricket duties. It was all very quick, if not abrupt, without the chance for a proper farewell – so these few paragraphs will have to suffice.

Why did Lane 'work' for me? I'm not sure. Like most people, I suspect, I'm quicker to object to what I dislike in sports commentary than approve of what I like. But some qualities always stood out with Lane, not least his understanding of cricket as serious fun.

Much of the fun in sport, I've always thought, is in the seriousness with which we treat it. Despite its loyal following, for instance, I've never warmed to England's Test Match Special: with their gravity-defying high spirits, they always sound to me like a bunch of drunks. It's considered a compliment to commentators now that they show 'the enthusiasm of a fan'. Yet surely if listeners want to partake of the enthusiasm of fans, they can sit in the crowd. I want commentators, not kibbitzers: it was one of Lane's cardinal virtues that he was always one of the former, and that while keeping a place for the fun he honoured the occasions he broadcast.

The best commentators, of course, know their media. Think of Richie Benaud. Last weekend I had my feet up at my cricket club and was watching an ING Cup game. Steve Waugh straight drove dramatically for four, then in businesslike fashion signalled for a new bat: his old blade had split down the back. The stroke was nothing, though, to Benaud's droll accompaniment: 'Some will wonder that a bat could incur damage from a stroke of such purity.' That's soooooo Richie. He never simply restates what we see; he confines himself, coolly and economically, to adding to it.

Tim Lane was as different as radio is from television. His was the voice of summer on the move. It was the voice in your car, on the beach, in the backyard. It was the voice, too, when you were out there doing it, or trying to do it, yourself; when you came off at tea, ready for a cuppa and a smoke, and wanted to know the Test score.

Lane knows sport, but he also knows broadcasting. He wasn't a former great speaking from a wealth of experience at the top level, and didn't try to be; he was our proxy, and understood that as his duty. He did the basics so well, in fact, that we tended to forget they aren't so basic. I recently tuned in to the ABC for an afternoon session of the Sydney Test against Zimbabwe, and listened for more than an hour without learning what had happened in the morning; mail bags, quizzes, the commentators' social lives and Kerry O'Keeffe's sophomoric jokes predominated. Lane was keen in his observations and artful with his language, but to recall his career by enumerating his various *bon mots* would do him an injustice. I liked Lane's commentary for the simple reason that he told me what was happening.

It would be disingenuous of me not to reveal a personal dimension to this note. I met Tim about thirteen years ago, have formed a friendship with him, and esteem him as a man as well as a commentator. But I mention this not merely as a disclaimer. I hope it won't embarrass Tim if I say that, at least initially, I found something about conversations with him a little odd. This puzzled me, as their content was always interesting and their tone always warm. Only gradually did I work out that it was *that* voice. I was so used to it coming from a radio – sharing an impression, giving a score – that it seemed all wrong to have to answer back. I'm over this feeling now. Indeed, it's enriched my appreciation of the Lane way. It is the essence of his craft that he doesn't have a commentary voice and a conversational voice: one is the other. Commentary is in his nature, and his nature is in his commentary.

Inside Edge Australia v India 2003–04

Ramachandra Guha
East Is East

'WHAT DO THEY KNOW OF CRICKET who only cricket know?' In the game's literature, this paraphrase of Kipling is perhaps the most famous challenge, posed forty years ago by the West Indian political philosopher C.L.R. James in his *Beyond a Boundary*.

What's striking is not how many have heeded the call to arms but how few. For the most part, cricket writing remains firmly in the cliché factory, a wholly-owned subsidiary of the sporting-industrial complex. Only occasionally is one reminded of James' injunction – such as while reading almost anything by Ramachandra Guha.

Guha has heard himself compared to James, and doesn't like it. 'I have heard it said that you should only read second-rate writers when you're young,' he says. 'That if you read Dickens, for instance, while growing up, you will never be able to develop a writing style of your own. I was lucky, I think, not to read James until a little later in my life.'

There are, however, decided parallels, in that Guha's writing about cricket is the richer for being a relatively small part of a larger oeuvre of social history. Just as the bulk of James' work bore titles like *World Revolution* and *Notes on Dialectics*, Guha began his career between covers by turning a Ph.D on the Chipko movement into *The Unquiet Woods* (1989), while his most celebrated book in India was an elegant biography of anthropologist Verrier Elwin, *Savaging the Civilised* (1998). When he refers to Tendulkar, he could just as easily be alluding to D.G. Tendulkar's multi-volume life of Gandhi.

Born in the north Indian town of Dehradun in 1958, Guha had the ideal first acquaintance with cricket for one destined to write of it: he both loved the game and wasn't very good. He developed an early affinity for such English writers as Sir Neville Cardus and A.A.

Thomson, while at school he bowled coarse off-spin and propped up the order from below.

Guha was largely lost to cricket in his twenties when, taking Proudhon at his word, he sold his hard-won library for a pittance. But Marxism proved inimical to a dawning interest in the environment: 'I was writing about Gandhi and Marx and ecology, but to the Marxists the ecology was simply a bourgeois distraction from the class struggle. I did not agree. Environmentalism might in the West have been about saving some pretty trees, but for the poor of India it was about their crops, the quality of their water, and access to grazing land for their subsistence needs. So the Marxists became disenchanted with me and I became disenchanted with the Marxists.'

Guha reconsidered his cricket apostasy in Australia fifteen years ago when he took advantage of a break in a La Trobe University seminar to tour the MCG. One of its blue-blazered guides recommended cricket bookseller Roger Page, from whose teeming shelves he soon gathered an armful of volumes.

Another incitement was the 1989 publication of Ashis Nandy's *The Tao of Cricket*, a much-discussed detour into the game by the eminent West-baiting sociologist. 'Nandy is a marvellous social analyst and critic of Indian culture whom I deeply admired,' says Guha. 'But his book was appallingly wrong-headed. He had a model of the world and he had fitted history into it ignoring everything that might conflict with his views.'

Provoked, Guha wrote *Wickets in the East* (1992), an enchanting 'anecdotal history' in the guise of a selection of all-time XIs for each Indian state, and *Spin and Other Turns* (1994), an exquisite evocation of the recent past sketching the heroes of Indian cricket in the 1970s. But his crowning achievement looks like being *A Corner of a Foreign Field* (2002), a fully-realised thematic account of the rise of the game in India with, *pace* Nandy, a caution against romance: 'There was no uncontaminated past when the result of a cricket match was not interpreted in ethnic or religious terms.'

Guha initially intended a biography of Palwankar Baloo, an untouchable who by his skill as a left-arm spinner at the turn of the century transcended the constraints of his caste. Tugging at this sporting thread, however, alerted him to a whole political and cultural tapestry: Palwankar's roles as an ally and later rival of the great untouchable leader B.R. Ambedkar, for whom he brokered the Poona Pact with Gandhi. Guha marvelled, as he often does, at how much historians do not know about India. 'India is simply the most interesting country in the world, period,' he says. 'It is also, often, cruel, corrupt and barbaric. Yet one cannot as a historian complain too much about this. Because in what other country in the world would you have so many themes to write about?'

His analysis spread in four directions – race, caste, religion and nation – and swept up a host of characters, some of whom emerge rather differently to their usual depictions. Lord Harris, the English cricket grandee whom history has cast as the father of Indian cricket, turns out to have been as insular and intolerant as most other governors of Bombay. Douglas Jardine, villain of Bodyline, is seen reunited in the land of his birth with his father's old native butler, tenderly accompanying him to the graves of relatives, then speeding him to hospital when the old man's strength falters.

The success of *A Corner of a Foreign Field,* which culminated in its recognition by England's Cricket Society as Book of the Year, was not foretold. Writing about cricket in India, Guha explains, is less a sinecure than one might imagine: his other non-fiction titles, indeed, comfortably outsell those concerning cricket. 'The intersection between the people who are interested in cricket and those who read books is actually very small,' he says. 'The book-buying public are not as a rule interested in cricket, even non-fiction, and the cricket-loving types don't buy books. So I like to quote that remark of Borges: "I write for myself and my friends and to ease the passing of time."'

Guha comes to the Sydney Writers' Festival promoting his cricket book, moreover, at a time when he is not writing about cricket at all. His latest project is a history of India since independence, his latest public role a skirmish with Arundhati Roy, the Booker Prize-winning novelist turned political incendiary who lent her name and profile to Andolan resistors of the Sardar Sarovar dam. In earlier days, the pair might have shared soapboxes; in these, when he has grown increasingly impatient with the stridency of Green politics, Guha described Roy in his book of essays *A Marxist Among the Anthropologists* (2001) as 'the Arun Shourie of the Left'.

'The extremists on one's own side are more dangerous than those on the opposite,' Guha philosophises. 'Roy could have performed a very valuable function of stirring an indifferent and somnolent elite to considering some issues of great importance, because success in the West matters in India. But she has now moved on to being an anti-war activist, and it is tempting to think that with every month there is a new issue: it's religion, it's Pakistan, it's George Bush.' It is one thing to have range as a thinker, believes Guha, another to have breadth.

The Bulletin April 2003

Wisden

Size Matters

ANOTHER DAMN, THICK, SQUARE BOOK, eh, Mr Engel? Scribble scribble scribble. Actually, the 136th edition of *Wisden Cricketers' Almanack* this year hits the doorstep with an unusually portentous thud, being the first to tip the scales at more than 1500 pages. One thousand five hundred and twenty to be precise. Wider still and wider/Shall thy bounds be set?

At first glance, this appears too much of a good thing. An annual of these dimensions seems to fly in the face of its editor's noble intent to 'make the *Almanack* less intimidating'. It may even be dangerous: I remember a story about a Jewish theologian who met his end when a copy of the Talmud fell on his head from a tall bookshelf. On the other hand, Mr Engel has my sympathy. Nothing testifies so eloquently to the preposterous congestion of international cricket schedules, and the surfeit of the game's abbreviated form, than the chunk of the *Almanack* dedicated to one-day internationals from Nairobi to Toronto. So I'll blame the game's administrators. After all, that's what they're there for.

At second glance, the annual's displacement seems to make reviewing it redundant. One might as well review the Albert Memorial or the Forth Bridge. *Wisden*'s contents are largely determined by events, so there's little point protesting about failures of character or plot. Were this a work of imagination, I might be tut-tutting: 'Mr Engel stretches the credulity of the reader by having England lose so many Test matches.' As it is a journal of record, one can merely pick bones with the soundness of its reportage, the authority of its judgements and the presentation of its information.

The present editor is seven years into his tenure, and I think has accomplished much. *Wisden* is more accessible, more structurally coherent, and contains more diversions than ever. It continues, moreover, to perform its traditional duties well. For me, the annual's vertebra is its treatment of English first-class cricket, and this remains not only meticulous but also flavoursome. The *Almanack*'s reporters have this year found space for the inability of Warwickshire substitute Darren Altree to find The Parks during his team's game against Oxford University, the Michael J. Powells fielded by Warwickshire and Glamorgan when they met at Edgbaston, Mushtaq Ahmed being trussed to a chair on the pitch at Bath in retaliation for his practical jokes while Somerset played Essex, and the ice-lolly break at Maidstone when Kent hosted Yorkshire. To borrow from the argot of management consultancy, this is 'value-adding' at its best, and something at which the annual continues to excel.

Engel has also striven to enrich the editorial matter that prefaces the time-honoured statistician's beanfeast. If this edition contains nothing quite as memorable as has been the case in recent seasons, there is again a good deal here to enjoy, especially Peter Roebuck's thoughtful appreciation of Graeme Hick and Allan Massie's delightful reflections on Scottish cricket. Engel's own notes are as engaging as ever, though I feel he lets Arjuna Ranatunga off rather lightly for his fifteen minutes of infamy at Adelaide during the recent Carlton & United Series (umpire Emerson's actions may have been foolish, but it seemed to me the height of bad faith on the part of the Sri Lankans to first invoke the ICC's opinion of the legitimacy of Muttiah Muralitharan's action, then to dispute an ICC referee's disciplinary jurisdiction). The general quality of the coverage is sustained throughout, perhaps the neatest line being Steven Lynch's description of the captaincy contest between Mark Taylor and Hansie Cronje as reminiscent of 'Dr Who outwitting the Daleks'.

The bones I might pick with this edition pertain to presentation. One development it is hard to like is the commercial infiltration of the *Almanack*'s pictorial section: this year's fourteen pages of colour plates are interspersed with eight gaudy and distracting colour advertisements. Was this not the annual whose editor in 1992 deplored

TCCB's decision to allow Cornhill logos on Test match outfields – which, by comparison, were positively tasteful – as a 'symbol of the level to which English cricket has to go to earn a crust'?

I'm also ambivalent, in an anorakish sort of way, about the degree to which *Wisden* has begun creeping from its accepted 'time zone'. It used to be that the end of the English season descended like a guillotine. Now it comes down like a perforator. In this edition, Ashes records have been updated to include the recently completed series down under. This introduces a loose screw to Bill Frindall's tight-wound mechanism: Pakistan–Australia figures, for instance, have not been recalibrated to include the series there that predated the Ashes. If you intend pushing the envelope, I think, it's best to push all the way or not at all.

It's impossible to greet the arrival of any *Wisden*, however, without experiencing a subtle lift of spirit. Work ceased for several days at my home when this edition arrived, and I read it as always as if in the grip of some cliff-hanging thriller. Unputdownable. And hopefully some years will elapse yet before it becomes unpickupable.

Wisden Cricket Monthly May 1999

Peter McFarline
Journo Man

ONE LAY DAY IN SYDNEY on England's 1978–79 Ashes tour, there was a knock on journalist Peter McFarline's hotel-room door. Outside stood Derek Randall, pen and paper in hand, evidently at a loss to complete a letter home. 'Ere,' said Randall. 'You're a bluddy writer. You write 'ome for me and tell the wife what's bin 'appenin'.'

As well as adding to the stock of whimsicalities regarding Randall, the story reveals something about the affection and esteem with which McFarline at his peak was regarded. *The Age*, Melbourne's broadsheet morning daily, has been blessed with fine cricket writers for a century, from H.W. Hedley, Frank Mauger and Percy Beames to Mike Coward, Greg Baum and Mark Ray in more recent years. McFarline, although he began his working life at the Brisbane *Courier-Mail* and in the mid-1980s worked as Washington correspondent for Melbourne's *Herald*, stands comparison with any of them.

McFarline's boyhood hero in Queensland was Ken Mackay: his reverence, he recalled, extended to mimicking Mackay's bent-kneed gait and the 'bovine dedication' with which he masticated his chewing gum. In one of his warmest pieces, McFarline described Mackay as 'facing up to the world's best bowlers in a manner rather reminiscent of W.C. Fields wielding a croquet mallet'. The adoption of Mackay as a favourite was a patriotic choice to be sure, but perhaps as much to do with a precocious sense of the role of personality in sport. McFarline always seemed most at home in his writing with characters, larrikins, rough diamonds – and less happy when the ranks of characters began to thin with the onset of professionalism. He revelled in sport's salty humanity, and lamented its steady leeching by money, media and managers.

In the case of cricket's professional revolution, of course, McFarline could claim to have been present at the creation: his best-remembered coup will always be revealing Kerry Packer's inchoate plans to form a breakaway cricket circuit. This May 1977 story reverberated round the world, and became part of *Age* legend (though the sports editor at the time, *mirabile dictu*, actually consigned McFarline's first Packer despatch to an inside back page).

At the time, McFarline was in his pomp, a tough, knockabout figure who socialised easily with a tough, knockabout Australian team under the captaincy of consecutive Chappells – indeed, McFarline had ghost written Ian's 1976 autobiography *Chappelli*. But the factionalised and poisonous atmosphere of that 1977 tour, I suspect, disillusioned McFarline at least a little. In Australia's *Cricketer*, he contrasted the trip with happier ones of yore: 'Little snippets, like a piece of criticism relayed from back home, in days gone by would have brought the rejoinder "have a drink, you bastard, what's this rubbish you've been writing about me". Oh no, not in 1977. Then it meant snarls, insults, a heated two-hour verbal or at the worst, no words at all. No drinks either.'

McFarline sided strongly with the Australian Cricket Board throughout the schism, although he was too sane a journalist to become an establishment mouthpiece or polemicist, as his book *A Game Divided* shows: it is lean, spare and punchy in the best traditions of Australian sporting reportage. But because he held no brief for anyone, he never hesitated to chide any athlete whom he saw as uppity, selfish or greedy; his judgements could be crushing.

McFarline's last twenty years were blighted by syringomyelia, a degenerative disease after an operation to remove a growth in his spine in 1982. The condition wrought a devastating toll on McFarline's health and appearance – he was wheelchair-bound for fifteen years, spent his last seven years in and out of hospital, and evolved a technique of first dictating then silently mouthing his articles to his unfailingly resilient wife Dell. Indeed, he was still appearing in *The Age* as little as two weeks ago.

McFarline's determination was such that he persisted for years on a biography that he had promised to write of the Australian Rules football guru Ron Barassi when there seemed at times little likelihood that he would live to complete it (it is dedicated, in fact, to Dell 'without whom there would be no book and no biographer'). Biographies of Australian sportsmen tend to be of either the cut-and-paste or tongue-in-bum variety. *Barassi: The Life Behind the Legend*, published in 1995, is by contrast painstakingly researched and scrupulously even-handed. McFarline had known Barassi for twenty-five years, hunted down scores of Barassi's friends and colleagues, and interviewed his subject for more than a hundred hours: the book as a result is remarkably candid and personal.

It is McFarline's best work, having been made by illness his most arduous task, and of a calibre that few Australian sportswriters today could approach, because it is based on a sort of simpatico between author and athlete that no longer seems feasible. No, they don't make writers like Peter McFarline these days, any more than they make cricketers like Derek Randall.

Wisden.com April 2002

C.L.R. James
Boundaries

IN IAN BURUMA'S 1992 NOVEL *Playing the Game,* the narrator makes a
pilgrimage to a distinguished nonagenarian critic cum cricket writer
from Trinidad. K.C. Lewis, now almost blind in his South London
retreat, is nonetheless surrounded by books: European history, Third
World politics, Renaissance art, English literature. The collected Marx
makes up one pile; a yellow column of *Wisden Cricketers' Almanack*
makes up another. In Lewis's intellectual panorama, delineated in 'the
cultivated accent of an English gentleman', Marx, Lenin, Shakespeare,
Milton and W.G. Grace are effortlessly elided. 'Do you know as a
young man I read *Vanity Fair* at least twenty times?' he asks. 'Cricket,
history and literature were part of the same thing to me.'

The *dramatis personae* of V.S. Naipaul's *A Way in the World,*
published two years later, features the West Indian Lebrun – an 'impresario
of revolution' who becomes the toast of literary and political
salons. 'It was rhetoric, of course,' observes the narrator. 'And of
course, it was loaded in his favour. He couldn't be interrupted, like
royalty, he raised all the topics, and he would have been a master of all
the topics he raised.' Lebrun is, it emerges, a figure of division in the
Caribbean: highly influential, utterly distrusted, with 'no popular
following', but capable of manipulating the mighty with 'big technical-
sounding words'. A politician discovering Lebrun's seditious streak
remarks ruefully: 'The man wants to take you over.'

It will come as no surprise to know that both Lewis and Lebrun
are thinly veiled representations of real people – they are almost too
vivid to be otherwise. But the trick of Buruma's Trinidadian Tiresias
and Naipaul's Caribbean Cassius is that they share common ancestry:
both were based on the West Indian C.L.R. James. His is a unique

reputation. Cricket lovers who celebrate his masterpiece *Beyond a Boundary* – which Warren Susman has called 'the most important sports book of our time' – usually have little sense of the protean political ponderings of which James' work is largely composed. Marxists and black studies scholars saluting him as an anti-colonialist avatar – Edward Said has honoured James as 'the father of West Indian writing' – have generally found *Beyond a Boundary* obscure, even unintelligible. In my well-thumbed 1983 edition, published in New York, four prefatory pages purport to describe cricket for Americans; they evoke C.P. Snow's remark that compared with explaining cricket to a foreigner, Chomsky's generational grammar is a snap. Few other thinkers can have cultivated such discrete constituencies; it is as though F.A. Hayek also wrote the definitive masterpiece about fly-fishing.

Because it is such a personal book, tracing the ideation of *Beyond a Boundary* requires some conversance with the details of James' life, in particular his education as a scholarship boy at Port-of-Spain's Queen's Royal College between 1910 and 1918. QRC drew on the English public school tradition embodied in Thomas Arnold's Rugby, whose ur-text is Thomas Hughes' *Tom Brown's Schooldays,* soaked in the ethos of fair play, restraint, perseverance, responsibility, and 'the old school tie'. Oxbridge-educated masters taught classics and history in English and European traditions. This educational and ethical conditioning 'never left me', James said, fostering abiding loves of cricket and literature. W.G. Grace's statistics and Shakespeare's sonnets did not compete for his attention; both were learned like passages of scripture. To everything he did, James brought a bookish intensity. According to biographer Paul Buhle, sixteen-year-old James responded to losing his virginity by reading all six volumes of Havelock Ellis' *The Psychology of Sex.*

In one of his famous letters to Felice, Kafka disavowed 'literary interest', describing himself instead as 'made of literature'. The phrase also suits James. The first signs he exhibited of political conscience are the short stories of his mid-twenties, examples of the Caribbean literary sub-genre called 'barrackyard fiction' (essentially, stories featuring working-class or ghetto characters). The inspiration was, however, anything but indigenous. James' characters could just as easily have been the drifters and low-lifers populating the stories of O. Henry, and his influences were the likes of Kipling and Maupassant. Political causes attracted him; James was gradually enamoured of the Trinidad Workingmen's Association, a group of militant returned servicemen making their presence felt in island politics, and impressed by their charismatic nationalist leader Captain Arthur Cipriani, who sought *inter alia* a minimum wage and an end to child labour. But it was his literary ambitions rather than his ideological consciousness that motivated James' pilgrimage to London in March 1932.

James arrived with three objectives: to publish his first novel; to ghost-write the autobiography of the great West Indian cricketer Learie Constantine, an old friend then playing league cricket in the Lancashire textile town of Nelson; and to complete an admiring study of Cipriani. The last and least likely actually came first: Constantine generously underwrote a small printing of *The Life of Captain Cipriani* in September 1932, which led to an English edition, *The Case for West Indian Self-Government* in April 1933. And the relationship between Constantine and James was unusual. It was actually not the intellectual wakening the athlete to issues of race; rather it was James who, at least at first, often found Constantine 'unduly coloured by national and racial considerations'. Nor was it James who brought radical politics to Nelson; rather was it Nelson, a stronghold of the socialist Independent Labour Party, that first exposed James to shafts of Marxist thought. Recently disaffiliated from the Labour Party, the ILP was in the vanguard of strike action by weavers in Nelson in August and September 1932, which James watched with dawning appreciation. Bliss it was to be on the left in that dawn of the United Front, the Left Book Club, *The Road to Wigan Pier*, POUM and the International Brigades, and to be a Trotskyite was sheer heaven. The exiled Trotsky was then a pin-up of Britain's liberal intelligentsia – his unsuccessful asylum application had been supported by George Bernard Shaw, J.M. Keynes and H.G. Wells – and James devoured Trotsky's *History of the Russian*

Revolution long before troubling to delve into Marx, relying in political debates on what his biographer Farrukh Dhondy calls 'intelligent bluff'. In retrospect, in fact, one sometimes wonders how deep James' political convictions *were*, at least initially, and whether their adoption was not actually part of his acculturation, an act of intellectual mimicry, like his assimilation of the ethos of fair play. Publisher Frederick Warburg observed in his autobiography *An Occupation for Gentlemen* that, far from affecting Orwell-like austerity, James 'enjoyed the fleshpots of capitalism, fine cooking, fine clothes, furniture, and beautiful women, without a trace of the guilty remorse to be expected from a seasoned warrior of the class war'. Like St Augustine, perhaps, James sought chastity and continence – but not yet.

All the same, James prospered between March 1932 and November 1938, the breadth of his output being as impressive as its volume. His original ambitions were all fulfilled. James' rendering of *Cricket and I* by Learie Constantine, though basic in its intent and sometimes breezy in its inaccuracy, was something of a landmark; as James commented, it represented 'the first book ever published in England by a world-famous West Indian writing as a West Indian about events in the West Indies'. *Minty Alley* then became the first novel by a West Indian published in the UK, and *Toussaint L'Ouverture* the first play by a West Indian to reach an English audience. But it was the historical events that inspired the latter which precipitated James' most original work: *The Black Jacobins*, a history of the slave revolt in the French colony of San Domingo that began in August 1791 and the twelve-year struggle to establish the state of Haiti, which James perceived as the hinge event in West Indian history.

Linking the slaves of San Domingo with the *sans-culottes* of Paris, James began a longer-term project to align the struggles of black and colonial peoples with European revolutionary traditions – with varying degrees of success. But in seeing European events as impinging on Haitian life, rather than vice versa, it is a genuinely path-finding work of anti-colonialist historiography. It relates a stirring

story ('One of the great epics of revolutionary struggle and achievement') and locates it firmly in the Caribbean as an outcome of sugar
and slave economics ('It is an original pattern, not European, not
African, not part of the American main, not native in any conceivable sense of that word, but West Indian *sui generis* with no parallel
anywhere else'). At the same time, it didn't elude James that *The
Black Jacobins* would chime with the times, composed as it was not
only as Italy warred in Abyssinia, but with 'the booming of Franco's
heavy artillery, the rattle of Stalin's firing squads and the fierce shrill
turmoil of the revolutionary movement striving for clarity and
influence'.

James' successes were not as a political seer. His survey of the
Third International, *World Revolution,* is possessed of fierce conviction and touching naivety: 'The huge fabrication of lies and slander
against Trotsky and Trotskyism in Russia will tumble to the ground,
and Stalin and Stalinism will face the masses inside and outside
Russia naked.' And as a fomenter of revolution himself, James was
also a failure. During fifteen years in the United States, where he
moved in November 1938 seeking further fusion between the racial
and revolutionary trends in his thought, he churned out agitprop
with the best of them. But the American left of the time was hope-
lessly fissured and ineffectual, a veritable People's Front of Judea, and
James' most significant work in this time wasn't at all what he considered the 'serious' stuff – books of the numbing obscurity of *Notes
on Dialectics* (1948) and *State Capitalism and World Revolution*
(1950). As he would admit, he was 'increasingly aware of large areas
of human existence that my history and my politics did not seem to
cover'.

Like Freud, James did not so much hate the United States as regret
it. Trying to understand why American workers remained deaf to his
dogma, James immersed himself in their popular culture, especially
the cinema, but also detective thrillers, radio plays and comics. Eight
long essays published posthumously in *American Civilisation* (1993)
suggest an energetic engagement with plebeian taste as well as a frustration with it. 'I remember that for years I pertinaciously read comic

strips unable to see what Americans saw in them,' he recalled. 'I persisted until at last today I am walking blocks to get my comics. In
Europe and when I first came here I went to see movies of international reputation. Now I am a neighbourhood man and I prefer to
see B gangster pictures than the latest examples of cinema art.' James'
interest in the politics of popular cultural forms also inspired
Mariners, Castaways and Renegades – a reading of *Moby Dick* as an
allegory of capitalism exalting Melville as the only 'representative
writer of industrial civilisation', written as chance would have it
while James was on Ellis Island awaiting deportation as an illegal
over-stayer.

In sport, James found a culture even more truly demotic. His final
quarrel with Trotskyism arose ostensibly because the movement was
indifferent to race, and cleaved to the idea of a vanguard party. But
James was vexed almost as much by Trotsky's dismissal of sport as a
distraction from the class struggle. Observing American mores,
meanwhile, had encouraged James to reconsider his own: 'I thought
constantly of cricket because I could not see it. I was constantly
thinking about cricket in this foreign environment. It gave cricket an
existence of its own with the elements of a beginning, middle and an
end. Whence this volume.' That is, *Beyond a Boundary*. Its composition began when James returned to England in July 1953, and James
was sending chapters to political allies in the United States by March
1957 which he challenged them to read 'without at a certain stage
being moved to tears'.

What is *Beyond a Boundary*? Like cricket itself, its kaleidoscope
of history, philosophy, literature and memoir can be difficult to
describe. It embarks from a poignant image: the junior James gaz-
ing wistfully through the window of his home upon the cricketers
at the nearby recreation ground. From here, he recalls, he could
also 'stretch a groping hand for the books on top of the wardrobe',
and 'thus the early pattern of my life was set'. Books provided intellectual sustenance, sport an ethical education: 'I never cheated. I
never appealed for a decision unless I thought a batsman was out,
I never argued with the umpire ... From the eight years of school

life this code became the moral framework of my existence. It never left me.'

A lengthy reconsideration of the author's youth ensues, with vivid depictions of famous cricketers such as Constantine and George Headley, alongside lesser lights like Wilton St Hill and George John, seasoned by the realisation that 'cricket plunged me into politics long before I was aware of it'. A key decision, for example, concerned which of two local cricket clubs to join, Maple or Shannon. Maple was a bourgeois, brown-skinned outfit; the people that as an educated man he would 'meet in life'. Shannon, noted for the 'spirit and relentlessness' of its play, was palpably of 'the black lower-middle-class'. James' choice of the former he regarded in *Beyond a Boundary* with anguished hindsight: 'So it was that I became one of those dark men whose surest sign of ... having arrived is the fact that he keeps company with people lighter in complexion than himself ... Faced with the fundamental divisions in the island, I had gone to the right and, by cutting myself off from the popular side, delayed my political development for years.'

At first glance, an affinity for radical politics sits uneasily with an abiding love of cricket. Henry Hyndman may have turned to Karl Marx because his omission from the 1864 Cambridge University XI rankled so deeply, but cricket's strong senses of tradition and hierarchy seem likelier to inculcate respect for established forms than desire to topple them. Nonetheless, just as James had sensed in *Moby Dick* the tensions inherent in capitalism, so in cricket had he located a metaphor for colonialism, especially in the symbolism that West Indian cricket teams were led always by members of the local white autarchy. Which made cricket a citadel worth storming – precisely because West Indian cricket teams taking the field under white leadership mirrored the exploitation of those who supported them, James sensed that a West Indian team with a black captain would be a powerful symbol of emancipation.

The second half of James' book considers cricket as history and art. James sees the great English cricketer W.G. Grace in Hegelian terms, as 'one of those men in whom the characteristics of life as lived by many generations seemed to meet for the last, in a complete and perfectly blended whole'. The term 'Eminent Victorian' applies to Grace, James feels, without Stracheyan irony: 'No other age I know of would have been able to give him the opportunities the Victorian Age gave him. No other age would have been able to profit so much by him.' James proclaims, above all, that the temper of the times expresses itself through sport, and that sport can only be seen in terms of those times. Paraphrasing Kipling, he issues his timeless admonition: 'What do they know of cricket who only cricket know? To answer involves ideas as well as facts.'

James thus accepts at face value what is often called the Golden Age of cricket – that period from 1894 to the First World War in which the values of amateurism are usually perceived as at their zenith, underpinned by both imperial confidence and a kind of prelapsarian innocence. The 'fall', as it were, is the Bodyline series of 1932–33, in which the English captain Douglas Jardine successfully undermined an upstart Australian team including its star batsman Donald Bradman through a policy of short-pitched fast bowling. James parallels the Jardinian cosh with the fascist jackboot: 'Bodyline was not an incident, it was not an accident, it was not a temporary aberration. It was the violence and ferocity of our age expressing itself in cricket.' Bradman, 'in his own way as tough as Jardine', subsequently presided over 'the age which can be called the age of J.M. Keynes'. This influence, characterised by 'the systematic refusal to take risks', still haunted the spiritless, passionless cricket of 1957, permeated as it was with 'the welfare state of mind'.

This, then, is James' historical schema. And had he published in 1957, *Beyond a Boundary* would have finished on a note of deep despondency. But another six years would elapse before its completion. In April 1958, after twenty-six years abroad, James was lured back to Port-of-Spain by a former QRC pupil, Dr Eric Williams, leader of the People's National Movement destined to lead Trinidad to independence. Enlisted to edit the party newspaper *The Nation*, and reunited with family, home and window, he began campaigning for the appointment of black Barbadian Frank Worrell to lead the

West Indies. Incumbent Gerry Alexander, a wicket-keeper and a Cambridge blue, was the latest in a long line of competent but unremarkable white cricketers to hold the captaincy; Worrell, with 3000 Test runs at an average of 53, was by far the Caribbean game's outstanding personality.

Like the character in George Moore, James had travelled the world in search of what he needed, then returned home in order to find it. How profoundly *The Nation* influenced the deliberations of the West Indian selectors is debatable: a sceptic might submit that Worrell, boy scout, university graduate and freemason, had the inside running all along. But, at the least, *The Nation*'s role in the conduct of the controversy entitled James a share in the successes that followed. Under Worrell, the West Indies participated in perhaps the most celebrated cricket series of the post-war period in Australia in 1960–61, won at the last gasp by the hosts, but in which the visitors obtained idolatrous support, culminating in a spontaneous parade down the streets of Melbourne. 'Clearing their way with bat and ball,' announced James, 'West Indians at that moment had made a public entry into the comity of nations.'

Evaluating James' importance in the rise of Worrell shouldn't obscure Worrell's importance in the rise of James. Not only did the summer of 1960–61 provide James with a logical and triumphal climax for his narrative, in which the West Indies were seen as sponsoring a sporting *risorgimento*, but it abetted his search for an English publisher. Hutchinson accepted *Beyond a Boundary* in June 1961, and published it in April 1963, on the eve of another West Indian triumph: a dramatic 3–1 victory over England on English soil. As the publication of *The Black Jacobins* had resonated with political upheaval at its release, so the launch of *Beyond a Boundary* seemed to echo the overthrow of sporting order. Here was West Indian cricket: united, uniting, irresistible. Here was James, *deus ex machina*, to explain it. Reviewers rhapsodised. John Arlott, the distinguished BBC broadcaster, proclaimed that 'in the intellectual sense, it is quite the "biggest book" about cricket or, probably, any other game, ever written'. It has existed in an echo chamber of endorsement ever since:

even Christopher Hitchens knows it as 'the best book ever published on the history and ethics of cricket'. E.P. Thompson proclaimed it the key to James' whole oeuvre: 'I am afraid that American theorists will not understand this, but the clue to everything [in James' work] lies in his proper appreciation of cricket.'

In its epigrammatic style and epic scope, *Beyond a Boundary* is assuredly remarkable. There had been nothing like it and, frankly, there has been nothing like it since. But it clicked with cricket and Caribbean readerships for reasons other than intrinsic merit. James flattered cricket's sense of its own importance with his view that 'aestheticians have scorned to take notice of popular sports and games' ('Good to see the eggheads waking up at last, eh?'), while his radical reputation cohered with cricket's sense of itself as a broad church ('That James feller's a Commo, dontcha know! Mind you, he knows his cricket').

James, meanwhile, offered his newly emancipated countrymen – for Trinidad achieved independence in August 1962 – access to a ready-made culture. Another famous expatriate whom Williams enticed back to Trinidad in this period was Naipaul, who unceremoniously bit the hand that fed him with *The Middle Passage*: 'I knew Trinidad to be unimportant, uncreative, cynical ... We lived in a society that denied itself heroes. History is built around achievement and creation. And nothing was created in the West Indies.' James elegantly demurred: 'What Vidia said about the West Indies was very true and very important. But what he left out was twice as true and four times as important.' At a Test match, James saw straightforward cultural substitution. Where the English had 'Drake and mighty Nelson, Shakespeare, Waterloo, the few who did so much for so many', the West Indies 'fill a huge gap in their consciousness and their needs' with the deeds of great cricketers.

Was it as simple as that? Probably not. James made a huge impact with the idea that sport had to be 'seen in its social context' – in sports history, to paraphrase William Harcourt, we're all Jamesians now. What James meant us to do, of course, was see West Indian cricket in *his* social context – the ethos of fair play that he imbibed at QRC.

For the work of such a radical intellect, *Beyond a Boundary* is conservative to the point of prudery. The *weltanschauung* of *The Black Jacobins* placing the colonial periphery at the centre of the narrative is nowhere evident. The understanding that the game is English remains unchallenged. Caribbean cricketers are inheritors of ancient English traditions which the English have grown too decadent to maintain – 'Thomas Arnold, Thomas Hughes and the Old Master himself would have recognised Frank Worrell as their boy' – rather than makers of new traditions. By depicting white West Indians as the forces of reaction, meantime, direct confrontation with the ideals of the colonising culture is avoided.

It is difficult to approach this without misgivings. The sense of incomplete and inherently uncompletable emancipation has left others deeply ambivalent about cricket's place in the Caribbean. 'Cricket is the game we love for it is the only game we play well, the only activity which gives us some international prestige,' wrote the novelist Orlando Patterson in his 1969 essay 'The Ritual of Cricket'. 'But it is the game, deep down, which we must hate – the game of the master. Hence it becomes on the symbolic level the English culture we have been forced to love, for it is the only real one we have, but the culture we must despise for what it has done to us, for what it has made of the hopeless cultural shambles, the incoherent social patchwork, that we have called Afro-Jamaican culture.'

James' historical analysis, meanwhile, is often tendentious. It is decidedly odd, for example, that with so lofty a conception of cricket he should invest so deeply in the eristic Grace. In *W.G. Grace: A Life*, Simon Rae points out that the great all-rounder 'had no truck with the emerging pieties associated with the phrase "it's not cricket"'. Grace 'never walked, never recalled a batsman even when he knew he should not have been given out, appealed with authoritative conviction from any part of the field, and whatever the game threw up would try to turn to his advantage'. What's especially curious about James' campaign to make an exemplar of Grace is that he knows this. He acknowledges, grudgingly, that 'it would be idle to discount the reputation he gained for trying to diddle umpires', then seeks to distinguish between what Grace went in for and *real* cheating: 'Everyone knows such men, whom you would trust with your life, your fortune and your sacred honour, but will peep at your cards when playing bridge at a penny a hundred.'

'Does everyone in fact know such men?' responded the historian Sir Derek Birley in *The Willow Wand*. 'And are they regularly entrusted with other people's fortunes?' The questions are fair. Grace certainly stands for something – just not what James would like him to. This seems an instance where James could not, as Chesterton said of Tennyson, 'think up to the height of his own towering style'.

It is the nature of Golden Ages that they are always in the past, and always irrecoverable. Cricket's Golden Age is no exception. In fact, it is something of a Potemkin village in cricket history: aspects of it were as gross and greedy as the mores of our own era. James misses it all: indeed, he wilfully looks the other way. His decision to define the period by its amateur ornaments, such as C.B. Fry and Archie MacLaren, is decidedly curious for a representative of both the left and of the imperial periphery. Where are the professionals? Where are the Australians? Where are the South Africans, the Americans, the Indian Parsees and, for that matter, the West Indians, all of whom toured England in the first decade of the twentieth century? Victor Trumper and Ranjitsinhji attract thirty mentions in *Beyond a Boundary* without their origins, in Sydney and in Jamnagar, being mentioned, or what this might mean. As for James' dismissal of the false prosperity of the twenties and the desiccated rationalism of the thirties as entailing 'the decline of the west', it suffices to observe that Gerald Howat, not only a very credible cricket historian but a biographer of Learie Constantine, has encompassed the same period in a book called *Cricket's Second Golden Age*.

The most serious strain on James' world view, though, is exerted by professionalism. Where *The Black Jacobins* is shaped by economics, *Beyond a Boundary* never mentions it. In particular, James ignores the paltry rewards earned historically by West Indian cricketers, up to and including Worrell's era, their need to play almost uninterruptedly in English leagues or for English counties in order to earn more than a subsistence wage, and the paradox that, in participating in the emancipation of West Indian cricket, team members had to condone their own continued economic exploitation.

In imagining his next world historical figure, too, James never foresaw an Australian whose *metier* was not playing ability but paying ability. West Indian players flocked to Kerry Packer when plans were revealed for his breakaway World Series Cricket in May 1977: they had no sense of 'the old school tie'. They evolved during those seasons a way of winning matches with unsparing speed and powerful batting: they drew, in other words, on the decadent 'age of Keynes', not on their gilded Caribbean antecedents, on the example of Bradman, not of Worrell. And James knew it. He was altogether nonplussed by the West Indies' twenty-year reign as the world's leading Test team; his only article about Clive Lloyd's all-conquering XI, in *Race Today* in October/November 1984, merely dismissed as 'nonsense' their claims to be regarded as great, managing in the process to sound like a blimpish colonel from Surrey.

James could be forgiven disillusionment. By the time *Beyond a Boundary* came out, he had actually returned to London, having fallen out, bitterly, with Williams over the presence of the US military at Chaguaramas, and the mooted West Indies Federation. James had strongly taken up the federal cause, becoming secretary of the West Indian Federal Labor Party. But, as Jan Morris observed, the region's islands were 'far less homogenous than they looked on the map and far more ambitious to be their own prime ministers'. The hostility of his reception when James attempted to re-enter Trinidad as a cricket correspondent of the *Observer* in 1965 spelt the end of his political ambitions there.

Henceforward, James' energies would be directed primarily to those regimes that sought his intellectual imprimatur, and he received political pilgrims with charm, grace and a sense of mild surprise – rather like his one-time hero Trotsky, who thought old age the most unexpected of all things that happen to a man. These, it must be said, make up a rather melancholy list, ranging from Kwame Nkrumah's corrupt Convention People's Party of Ghana to Maurice

Bishop's ragtag New Jewel Movement in Grenada – hence Naipaul's 'impresario of revolution' tag for 'Lebrun'.

'K.C. Lewis', as it were, has done rather better than 'Lebrun'. *Beyond a Boundary*'s reputation endures. It remains, nonetheless, a work of arresting and sometimes disconcerting paradoxes – in particular, that such a passionate apostle of upheaval should have so failed to factor its possibility into his favourite game.

New Writing 12 (2003)

Home and Away

Club Cricket
Naughton's Old Boys

THESE ARE DARK DAYS FOR THE game of cricket. As I write these lines, Hansie Cronje is in the dock, Salim Mailk in purdah, the International Cricket Council struggling to the eradicate the menace of match-fixing with all the finesse and subtlety of Inspector Clouseau. For the ordinary grass-roots cricketer, these developments are perplexing indeed. After all, we have been playing badly for years, and not once have we been offered a cent for it.

Jim Young's delightful memoir of his experiences as a grass-roots player gives a voice and a focus to this base of the cricket pyramid. The many thousands who donate their Saturdays and Sundays to the game, for no reason other than the pleasure of the companionship and the possibility of some shared suffering, are under-represented in the game's literature (perhaps we're wary of letting our wives and girlfriends in on what we're really up to). They lend cricket a special quality. As E.V. Lucas put it in *English Leaves*: 'There is no other game at which the confirmed duffer is so persistent and undepressed.'

I can make this judgement, I think, with some authority. At the age of thirty-four, I have been a humble – nay, very humble – park crick-eter for a quarter of a century on and off. Jim's beloved Naughton's Old Boys have an everyteam quality. I have played with and met char-acters very like them; if you have played for any length of time, you will have, too. They belong, moreover, to an honourable tradition. Certain passages of this book call to mind the famous Allahakbarries XI of a century ago, whose captain J.M. Barrie laid down such immortal commandments as: 'Don't practise on opponent's ground before match begins. This can only give them confidence.'

From anyone who has savoured the pleasures of park cricket, this book will stir reminiscences of one's own. I feel a particular personal pang of recognition: like Jim and his pals, I favoured the front bar at Naughton's; I played for several seasons at Edinburgh Gardens, whose prettiness Jim rightly attests; I can corroborate, too, how testing of character is the task of playing in a competition where one is called upon to umpire as well.

Not so long ago, I was taking my turn at this duty in a game in which I was involved, and enjoying the summer scene: the mellow late afternoon light, the enchanting birdsong, the resonance of willow on leather. But my reverie was disturbed by a sulphurous lbw appeal from the opposing captain, a purveyor of nude dibbley-dobblies. I suspected it was plumb, but suspect was all I could do, as my mind was momentarily as blank as Alan Bond's in a witness stand. Shaking my head, I stammered something about it 'doing too much'. The captain, unconvinced of the sincerity of my tribute to his trickiness, penetrated this as the excuse it was, remonstrating that I was a 'cheating c—'; not once, but for the balance of the day. He had my sympathy. Fortunately, for honour's sake, we lost by a suitably comfortable margin.

Some months later, in my capacities as a journalist, I was researching a story that required some expert testimony on an aspect of sporting commerce. It was recommended I consult one expert in particular, and duly made an interview appointment. Pad and tape recorder in hand, I was ushered into an office to face, yes, my chief tormentor of that day, the captain I had so grievously wronged. I conducted a curious interview, looking everywhere but at my subject, and with intense regard for my notes; had it been possible, I would have asked my questions while facing in the opposite direction. Finally, at the end of our chat, with no real reason for preserving my anonymity, I asked my interviewee if he recalled our previous encounter. Fortunately, once apprised of its circumstances, he proved generous and forgiving.

Ah, the healing power of time, I thought, and cricket's fraternal spirit. For a while anyway. I next set eyes on my new-found friend at the start of the following season, and extended my hand in greeting. 'Oh, it's you,' he replied, 'you cheating c—.'

I'll go no further, as this is Jim's story. But it does, in a sense, belong to all of us who've ever strapped on a pad (preferably with old-fashioned buckles), picked up a bat (with the maker's name unintelligible), or wondered why we play this slow, difficult, infuriating and utterly bewitching game.

Introduction to *Any Old Eleven* by Jim Young (2000)

Amateur Sport
Struggletown

SATURDAY EVENING AT MY CRICKET CLUB is always a festive and usually a comradely occasion. The day's events are relived, growing greater and worse with each round of drinks. Differences between opponents melt; shared experiences solidify.

After a while, it can usually be guaranteed, the talk will take a practical turn. How are you guys off for numbers? What are your pitches like? Yet the loudest common chord is usually struck by a description of one's local council as either passively indifferent or actively antagonistic where organised sport in their area is concerned.

Rising costs, declining services, pettifogging administrative requirements, and a generally sneering attitude are heaped on more general complaints like disappearing parks, dwindling numbers of umpires, increased public liability insurance and the weighty legal burdens on voluntary committee members. An office bearer of one club I spoke to recently, for example, told me of a local council meeting where two councillors had wondered openly why they had a local team at all: cricket, it was well known, was an elitist activity.

To be fair, councils tend to be proxies for blame that might well attach to others; it is merely that they are the principal point of contact, and thus abrasion, between private club and public policy. I can sympathise with council staff who must deal with a constantly changing cast of honorary presidents, secretaries and treasurers, each with differing degrees of diligence and aptitude. Community sporting clubs aren't usually known for high standards of corporate governance and clean lines of accountability.

Yet my impression, from personal experience as a volunteer committee member and years of anecdotal exchanges with others in the same racket, is that councils tend to regard local sporting concerns as either bastions of blokey boorishness, as civic amenities like adventure playgrounds, or as milch cows of revenue – an attitude in keeping with modern 'user-pays' principles of local government, and the dreary rationalism of our age.

Certainly, council amalgamations, austerities and outsourcing of services have had a massively deleterious affect on the quality of suburban and rural recreational fields. In May, John Thwaites' new Department of Victorian Communities was allocated $72 million for upgrades to community sport and recreation facilities across the state – a worthy initiative to be sure, but also a tacit admission of the extent of their deterioration, and ultimately a case of trying to reinvigorate what had been steadily bleeding away.

Sporting clubs, too, remain soft targets for technocrats trying to make budgets balance. Residents who resent their rate burden can, if it comes to that, up and leave. Clubs aren't gypsy caravans. They can't simply move on. Their whole identity is embedded in and inseparable from their location.

When a sporting club disappears it is usually a sign that something is amiss in that location. There is no more poignant index of the economic and social travails of regional Victoria than the approximately 130 country football teams that either folded, went into recess or merged in the twenty years to 2001.

This is troubling. Those who champion Victoria as a great sporting state and Melbourne as peculiarly passionate about sport usually have in mind marquee occasions, like the AFL Grand Final, Melbourne Cup, Boxing Day Test and Australian Open, and great arenas, from the Melbourne Cricket Ground at one end of town to its *nouveau riche* cousin at the other. But this no more makes us a great sporting society than our queuing to see the *Matrix* and *Lord of the Rings* trilogies make us a great movie society.

What should gladden all our hearts is that we remain a people that participates in sport, rather than consuming it passively as another entertainment industry option. Technological change has individualised and virtualised leisure, social change has deregulated working hours and gender roles. Yet unlike many industrial democracies, Australians still like doing it themselves: even with our ageing population, almost a third of males and a quarter of females over the age of eighteen are involved in some organised sport or physical activity.

So it is in Victoria. According to the Australian Sports Commission, this state accounts for a quarter of national recreational involvement in sport. In terms of proportional involvement we rank highest in tennis, second-highest in a group of sports as diverse as cricket, cycling and martial arts; the enjoyment of sport, indeed, is now perhaps more widely harnessed than ever.

A couple of weeks ago, while my club-mates and I sweated away in the practice nets at our suburban ground, I enjoyed a marvellous Australian *tranche de vie*. To our left, thirty young women, earnest enough to be in uniform, were involved in some vigorous touch football. On the far side of the oval, a pretty serious game of improvised soccer was in progress, witches hats for goalposts. A group of twenty runners, evidently members of a club, panted past. On a pitch in the middle of the ground, two fat guys bowled a tennis ball to one another; two young dudes nearby flung a frisbee; walkers, strollers and sun worshippers spectated. It was blazingly hot; the games had that peculiar mix of deep seriousness and bursting exuberance; amid all this unselfconscious physicality, I suspect I could have been in no other country in the world.

It is in order to touch the sporting nerve, and swing the sporting vote, that governments have invested so wholeheartedly in academies and institutes for the incubation of athletes of international standard. Thus, too, and rather less appealingly, their prostituting themselves for the sake of big events – often carving up the very parks providing the people's sport with its tenuous foothold in the community.

It's curious, in fact, that we have such a highly developed understanding of sporting excellence, acknowledging as we do that even the greatest natural ability needs coaching and resources to unlock, while remaining so blasé about the sport of the average man, woman and child, taking it for granted that the urge to involvement and the spirit needed for its organisation will endure come what may. It's curious too that governments believe so unstintingly in sport as a contributor to national pride, yet remain so indifferent to it as a source of social capital.

Now and again, of course, recreational sport gets a grudging pat on the head for contributing to public health: VicHealth is currently sponsoring a three-year, $25 million program aimed at lifting participation rates. This, however, is a reductive view: you could improve the quality of fitness in this state simply by paying people to jog on treadmills. What's of prime importance where recreational sport is concerned is the expression it gives to our instinct to associate, and the habits it instils of giving, sharing, volunteering, building culture, making tradition and joining in common purpose – customs of such immense value that they are beyond price.

It was Alexis de Tocqueville in *Democracy in America* (1830) who first perceived the connection between the strength of democracy and the vibrancy of civil association: 'Americans of all ages, all stations in life, and all types of disposition, are forever forming associations … of a thousand different types – religious, moral, serious, futile, very general and very limited, immensely large and very minute.'

When Melbourne's oldest and most venerable association was formed eight years later, it was for sport: the Melbourne Cricket Club. Victorians have tended to need pretty serious rousing to take up religious and moral causes, and we require the threat of a fine to make us vote, but we've seldom failed to rally in sport's name. The challenge – a stiff one in time of political expedience and economic exigency – will be to keep it that way.

The Sunday Age December 2003

The MCG
Place of Sense

LAST YEAR, ON A FLIGHT OF STAIRS at the Melbourne Cricket Ground, I encountered a long-lost family friend whom I'd not seen in the twenty-five intervening years since we'd watched the Centenary Test together at the same ground.

We exchanged hellos of pleasure but not of surprise. And it wasn't long before we were sharing a conversation probably not unlike our last. What about that knock of Gilchrist's, eh? What price those Cats this season? It was as though neither of us had actually left the ground in the interim; we'd just gotten separated, as you do, midst the crowds and clamour, for a quarter of a century.

Such is the nature of the MCG. Indeed, it is all our yesterdays – the sight not merely of Test matches, Grand Finals and an Olympics, but Highland gatherings, ascents by pioneering aviators, concerts, conscription rallies, military encampments, evangelical crusades, papal and royal visits. It has seen Sir Donald Bradman score nineteen first-class centuries, Wayne Carey kick 400 goals, Betty Cuthbert win three gold medals; it has heard Billy Hughes speak, Billy Graham preach and Madonna sing – or whatever it is she does. But more than that, it is the source of numberless personal moments, musings, meetings, sensations. In offering history to recall, it affords us glimpses of ourselves in the making; one helps navigate the other.

Megan Ponsford's work is not about sport per se; nor is it about history, though it is historic. It foregrounds the arena itself – or, to be more specific, the northern side of the MCG, just before the wreckers moved in to commence its top-to-bottom refurbishment. It is not the colosseum of cliché, of grand occasions hosted, of mighty deeds done; it is not the ground crammed to the bleachers, upholstered with advertising, viewed by television from reverent elevation. It is our MCG – that is, the MCG of the fans, of the faithful, even if, so artful and intimate is Ponsford's perspective, they may at first have trouble recognising it.

You'll recognise the name Ponsford at once. In the 1920s and 1930s, innings played by her grandfather Bill were like semi-permanent installations at the MCG. Twice he exceeded 400 there: the cricket equivalent of scaling Everest and K2. Bill Ponsford also joined the Melbourne Cricket Club staff in 1932, and was inter alia manager of the turnstiles in February 1961 when the ground hosted its Test record crowd of 90,800; it is said that he hid the sacks of pounds and pennies in the assistant secretary's office until the banks opened on Monday.

The naming of the W.H. Ponsford Stand in Bill's honour in 1967 left his grand-daughter with a deep attachment to the MCG. 'Generations of my family have been members of the Melbourne Cricket Club,' she says. 'My children are on the waiting list. My grand-father swapped his ladies' ticket for my membership in 1985 when I turned eighteen – the year the club admitted women. Obviously, I have always been very proud to have my family's name on such a prominent building and I wanted to have images of it to show my children and their children.' When the Stand fell last year and her family distributed the lettering, she acquired F and S for her boys Eduardo and Alesandro.

Yet Megan Ponsford does more here than stock a family album. It's arguable that the rebuilding of the MCG is more in keeping with its traditions than preservation. The ground has a habit of periodically devouring itself. Its past is strewn with stands first raised then razed: the Grey Smith, the Harrison, the Wardill, the Smokers. The new members' enclosure will be the fourth fulfilling that role. We have little, however, by which to remember what it was to be 'in the MCG', as distinct from 'at the MCG'. The weather-beaten wooden benches and bare concrete catacombs of the Ponsford and Olympic Stands might have no place in a future of contoured plastic comfort. But they are part of our past – a past already fast receding. Paradoxically, though few of Ponsford's images feature people, people are her chief concern: the MCG's fashioning to suit the customs and habits of the age.

I have a slightly unusual attachment to the MCG, having spent more time at the ground when it has been empty than full. My species of research has at times entailed almost semi-permanent occupancy of the library and club archives. Ascending the stairs of the members' stand, passing the prints on the walls from Chevallier Taylor's 'Cricketers of the Empire' series, was the loveliest 'commute' you can conceive of; the hum of quiet scholarship and the clink of morning tea cups in the library were warmly welcoming; the cases of bats, the images of touring XIs, the portraits of Melbourne Cricket Club secretaries could and have absorbed me for hours. And when I come to recall the sport I've watched at the ground, it comes back to me not in a catalogue of big names and great feats but in sideways glances, backwards looks, and momentary illuminations.

From the very first Tests I witnessed almost thirty years ago, I recall the sensation of waiting. Practices of patience do not come naturally; I'm sure cricket at the MCG helped inculcate mine. I grew to love the Boxing Day morning hush, the anticipation, the orders of precedence, the ceremonies. I relished, for example, the fuss of team photographs. In those pre-celebrity days, before they all hosted their own talk shows, one seldom saw cricketers except when playing. Yet here they were, all stiff reserve and ungainly pride, lining up like me and my schoolmates for class photos. From those days, I am sure, stems my deep distaste for the modern custom of practice on the arena, and my impatience with the daily pre-match promenade of commentators and bureaucrats in search of reflected limelight.

Another education was in the idea of scale. Most adult objects and ideas can be reduced to specifications suitable for children. Not the MCG: its dimensions were, and remain, uncompromisingly adult. Test matches were so vast, the hubbub so unending, the action so far away: I had to recalibrate my senses and my imagination, as it were, in order to fit it all in.

The first cricketer I saw up close was Doug Walters. My brother

and I were occupying the front row on the ground level of the
Ponsford Stand when Walters came haring to the third man boundary to retrieve. I'd never seen a man run like that, so far, so fast, so
urgently; I'd never guessed that actions could be as natural as his
stoop to capture the ball; and while nobody today remembers Walters
as an outstanding athlete or fielder, I swear he threw that orb almost
into orbit, before following it in at a nonchalant jog. Above all, I'd
never seen anyone so rapt in a physical task. We could almost have
touched him, my brother and I. We could see the crease of his flannels, the perspiration beaded on his face, the recession of his hairline;
we could sense the thud of his boots. Yet he never even sensed us, or
any of the proximate thousands. This was better than television and
movies; even theatrical actors didn't charge headlong towards you,
grab, swivel and hurl. I wonder now whether that was the instant
cricket grabbed me, and has yet to let me go.

While youthful sensations are outgrown, the MCG also provides
an abiding set of experiences: those of being in a crowd. The dynamics of the Australian sporting audience are unique. We're a big country thinly populated. Seldom do we feel the insistent press of others;
in almost no other context, sport without a crowd being unthinkable,
do we actively seek it. Personally, I'm not over-fond of big cities and
busy thoroughfares, or even too much company. But I relish forming
part of the MCG crush; nowhere else, I think, can one enjoy such a
perfect balance of solitude and companionship.

No-one in my memory has quite ignited a crowd like Gary Ablett.
My favourite recollections of him, in fact, are not of marks and goals,
but how it felt to be one of his audience. I remember, for example, a
night game between Geelong and Carlton, in which Ablett was for the
most part subdued by Steve Silvagni, but which at one point he lit
with a lightning bolt of brilliance. It started when a mis-kick rolled
into open country round half back. The only player in the neighbourhood, that dependable defender Michael Sexton, had only to
control it, and the clearance would be Carlton's. All at once, though,
the ball began bouncing treacherously, confounding his grasp, rolling
just out of reach. Again and again Sexton groped for it, unsuccessfully,

with the mounting frustration of a man scrabbling in the dark for a dropped key.

Spectators then became conscious of the approaching Ablett – not merely approaching either, but eating up the distance with ravenous strides. The anticipation was fearful; one could almost feel the force of the coming crunch of bodies. Yet it was, simultaneously, hilarious, pure pantomime; you expected any second a chorus of: 'Look out behind you …' When Sexton was engulfed in a tsunami of a tackle, the spectators unleashed their own wave, a collective exhalation ended in a rolling roar of laughter, appreciating both the unintended slapstick comedy of what they'd just witnessed, and their own inno-cent absorption in it, as though at the same time both being football fans, and watching themselves be football fans.

It could be argued that we lose little of architectural distinction in the northern side of the MCG. Its rugged, rough-hewn functionalism, its colliding styles and crenellated skyline, were not to everyone's taste. Nor could Megan Ponsford be accused of glamorising the ground, or even the stand bearing her grandfather's name. There is beauty here, but it is an austere beauty, a decaying grandeur, alongside some taw-driness, and the occasional weird flourish; the Long Room's plushness is contrasted with riots of concrete, stairwells, fences, poles and pipes.

Nonetheless, we are losing something of the MCG experience. The Ponsford and Olympic stands embodied principles of sport specta-torship on which we should place greater value: they accepted anyone and everyone, protected no privilege, needed no eye-catching gold card or platinum pass to enter. The old-fashioned wooden bench was a vestige of the understanding that spectators could and would share their space fairly; in the one-man, one-seat world, ostensibly more democratic, we are actually absolved of the expectation of reciprocity. Nick Hornby has remarked that English football, the self-styled 'people's game', is 'prey to all sorts of people who aren't, as it were, the people'. I'd hate to imagine this true of the MCG, the self-styled 'People's Ground'.

I wonder, too, how cosseted spectators need be. Television's perva-sion of our games seems to have instilled a belief that live sport should only be watched in lounge-room conditions; the creation of 'family-friendly environments' is a charter for arenas all over the world. Yet the phrase 'family' has acquired a number of unappealing undertones: its invocation by opportunist politicians, hugging the phrases of their focus groups, is especially creepy. I perceive a misunderstanding of how and why we watch sport. It isn't simply another leisure industry option. Even the superbox habitué seeks to partake of an experience different – more raw, less predictable – to that obtainable from other streams of corporate entertainment, if once or twice removed. Yet if present indicators are a guide, succeeding generations will scoff at our quaint habit of attending events in the open air. The past isn't something the MCG has ever sealed itself off from. I hope it never does.

Introduction to Megan Ponsford's *Home Ground* (2003)

Australians in England
At Home Away

PLAYING CRICKET ON A PATCH of 'rough-and-ready parkland' near Bondi Beach in the early 1920s, Jack Fingleton and his boyhood pals had a solemn ritual. Each week their little arena had to change name, and only an English ground would do: it became, in turn, Old Trafford, Headingley, Lord's, Bramall Lane, the Oval. There was, Fingleton explained, simply 'more romance' to these faraway arenas. Visiting them as a Test cricketer in 1938 and later as a journalist would duly feel like touring his own youthful imagination. 'The first step on London soil,' he felt, 'is the big moment in any Australian's life.'

It is hard to imagine any of Steve Waugh's Australians feeling similar stirrings today. England is just another cricket destination, one with which they are thoroughly familiar, not only as internationals but as county professionals. And though Waugh himself has loyally described England as 'the ultimate tour, steeped in tradition and history', he also fully favours this year's truncated schedule, squeezing five Tests into the span of fifty-four days: previous tours were 'too long anyway', tour matches 'a waste of time'. Not much romance there.

That's partly, of course, because of what is possible. Ships and trains having given way to coaches and planes, everything nowadays is faster. In a business-class seat on a Singapore Airlines 747, it took Waugh's men less time to travel to England than it took Billy Murdoch's 1880 Australians to proceed from Dewsbury to Belfast by carriage, train, paddle steamer and jaunting car. But it's also true that where cricket teams once undertook tours, they now take trips. Where Waugh's team will be away from Australia for a total of three months, Murdoch's men were gone almost fifteen months, their journey also bookended by games in Australia. The very first representative

282

Australian team 123 years ago took an even more serpentine route, playing in Brisbane, Toowoomba, Sydney and Maitland on the outward route and in New York, Philadelphia, Toronto, Montreal, Detroit and San Francisco on the homeward leg.

Yes, it's a different business, touring England today – as different, indeed, as Anglo-Australian relations. Of that first team, *Punch* gushed: 'Australia may well be proud of you, and England too, as branches of the British willow, though grown in the antipodes.' The idea of Australian cricketers as welcome pilgrims to the centre of race, culture and cricket is now an archaism, on both sides. This was brought home to Steve Waugh on his last tour when six louts outside a cinema leered at him: 'You know what you Aussies are going to win this summer? Fuck all, you Aussie bastard.'

For all the changes, however, there remain some surprising continuities. Behind both the brevity of the current visit and the elongation of the earliest expeditions, for example, lies a common cause: money. In 2001, it is a matter of maximising international playing days to optimise revenues. In the pioneer era before the creation of the Australian Cricket Board, more play meant more spoils to divide at tour's end; the original tours were mercantile adventures little less than sporting challenges, pride and profit concordant motivations. Managers were appointed by the players to look after their interests, and to carve up proceeds. 'We have some good days in England,' recalled Hugh Trumble, 'but the best of all is when we finish up.'

England, too, is where traditions took hold that have never been departed from. It was here, rather than in Australia, that the idea of a green and gold raiment took root. Joe Darling's 1899 tourists were the first to sport it, even flying a green and gold standard over their Holborn hostelry, the Inns of Court Hotel. Famously attached to their headgear, Steve Waugh and Justin Langer have a spiritual antecedent in Victor Trumper. 'Trumper always wore an old Australian XI cap,' reminisced his team-mate Clem Hill. 'It was bottle green, but nevertheless he stuck to it to the end, and there was no end of bother if Duff or some of the other humorists of the side got hold of the cap and hid it.'

In the tight-lipped professionalism of the current Australian ensemble, one also detects echoes of past XIs. A correspondent of *The Strand* who visited that team of Darling's described them in terms not ill-suited to Waugh's; while he could 'recommend a breakfast with the Australians as a first-rate recipe to anyone afflicted with an attack of the blues', he perceived the 'serious spirit' of their quest for victory, where 'anything likely to interfere with their attaining that result is to be rigidly eschewed'. A visitor from *Penny Magazine* six years later conscientiously enumerated breakfast habits with a flavour of the times: first course was always porridge, coffee scorned as 'bad for eye and nerves'. He also found the Australians complaining about the weather, Reg Duff observing drolly that 'the English summer has set in with its usual severity' – a complaint that has surely never altered.

As it has also vanished from much of English life, the pomp and circumstance of tours past is long gone. Bob Simpson recalled twenty speeches on his first twenty days in England as captain in 1964; Steve Waugh in his *Ashes Diary* for 1993 reported a consensus that 'most of the lads believe one function a week is too many'; the MCC dinner at Arundel is now the only one of any antiquity. In some ways, this public side of the tour has been supplanted by media duties. A century ago, touring cricketers willingly signed contracts that prohibited their speaking to journalists altogether: the practice was thought potentially divisive. The Board later imposed strict limits on player dealings with the press: forty years ago, captain Richie Benaud was 'gagged' by manager Syd Webb, an infamous martinet. Australian players writing about their doings are today commonplace: about half the current team are contracted to media organisations in 2001, while Justin Langer provides daily 'postcards' on the Board's own website.

In days of yore, from 1886 to the Second World War, Australian teams in England were accompanied by an honorary medical officer, Dr Rowley Pope. An eccentric ophthalmic surgeon from Sydney, he accompanied the tour at his own expense, doling out cures, keeping records, filling in as a fielder, and generally advising graceless antipodeans about English etiquette, like bowing a black tie, and recognising a fish knife; Arthur Mailey remembered him as 'a good Australian and, like all good Australians a good Britisher'. The team physiotherapist today performs services more scientific and special-ised; no need to recognise a fish knife in an age when the use of cut-lery often seems entirely optional.

In other senses, though, this tour will re-establish traditions in desuetude. As part of an embryonic 'player education' program, for instance, Waugh's Australians stopped *en route* to England to visit Gallipoli, where Australian troops in April 1915 were involved in some of the bloodiest and most futile actions of the First World War. This doffing of the baggy green to the slouch hat revives an old custom. Warwick Armstrong's Australians were the first to attend an Anzac Day ceremony in London, laying a wreath at the Cenotaph – a custom continued into the 1960s. The 1930 Australians passing through Paris even paid respects at the tomb of France's Unknown Warrior.

Another instance of going back to the future is in the attitude to wives and children. In the days when players ran their own tours, spouses were not uncommon: five members of the 1902 party, for instance, made the journey thus accompanied, Trumble taking the tour as his honeymoon. Not everyone approved – Monty Noble, Australia's captain in 1909, thought wives responsible for 'uneasiness and disquietude of mind in the team's family circle' – and there was satisfaction as well as chagrin when the Board stamped the practice out. A member of Bill Woodfull's 1930 Australians told journalist Geoffrey Tebbut: 'You can't beat women when it comes to starting trouble … No, it's bad luck for the wives, but it has either got to be a cricket tour or a large-scale tea-party, and you know the kind of trou-ble that is brewed over cups of tea.'

One team member at least would have demurred from that judge-ment. Donald Bradman found the separations from his wife Jessie in 1934 and 1938 so trying that he threatened to retire on the latter tour unless the Board granted her permission to join him at that trip's completion. Policies of varying liberality over the past twenty years have now finally turned the clock back a century. Under the terms of a four-year memorandum of understanding between the Board and the Australian Cricketers' Association reached in May, members of Steve Waugh's side will be entitled to bring their partners at any time. Those there for more than fifty days will have airfares and a fortnight's accommodation paid for. The captain – with two children and a third on the way – is the instigator: he believes 'that's the way it should be'.

The ways Australian cricketers amuse themselves in England have, naturally enough, changed with the times. Frank Laver appears in his *An Australian Cricketer on Tour* (1905) driving the stately Rover he ordered after a visit to that company's Coventry factory; Steve Waugh's *1997 Ashes Diary* finds him behind the wheel of a souped-up BMW at Brand's Hatch. Don't expect to see any of this mob at *Lohengrin*, like the Australians of 1899, or the 'Black and White Minstrel Show', like the Australians of 1964; Waugh's men prefer Bon Jovi. Nor would one anticipate them in the neighbourhood of the Rufus Stone in the New Forest (the 1921 team), the Savage Club (1938), the Bristol Brabazon (1948) or a potter's in Stoke-on-Trent (1972). In 1993, Merv Hughes confessed not to have heard of Stonehenge: comrades told him it was 'the pile of rocks that Chevy Chase knocked down in *National Lampoon's European Vacation*'.

Again, nonetheless, certain common recreations remain unaltered. This being an Australian team, a degree of drinking is to be expected in 2001: more than in 1930, when twelve of the team were abstainers and the feats of Bradman, Bill Woodfull and Clarrie Grimmett were celebrated by temperance movements, though less than in 1893, when a drunken brawl among players left blood all over a railway compartment, and in 1912, when raucous Australians were refused service by stewards on a voyage to Ireland.

There will, above all, be an awful lot of sport. Australian cricketers abroad are – and have always been – sports fans as well as sportsmen. Racecourses represent a particular enticement: members of the 1880 team went to Epsom for the Derby, where Fred Archer won on Bend Or. When the 1899 Australians were made members of the Manchester Racing Club, they made a killing on a horse called Portobello after overhearing a *sotto voce* tip.

Keith Miller was perhaps the most famous Australian turf devotee: his legendary 10/122 in the 1956 Lord's Test followed hot on an appearance at Royal Ascot, where he was described having given 'a glittering display of What the Well-Dressed Man Should Wear'. But Richie Benaud's Australians of 1961 were just as keen, four of them leasing a racehorse called Pall Mallan: when it won its fourth start, Barry Jarman's cry of 'you bloody beauty!' was heard all round Headingley. Sadly perhaps, cricketers in the company of bookmakers now occasion different thoughts. The carefree days of 1993 when Steve Waugh could document the sight of his brother in a Coral's tent at Leicester – where he 'backed horses all over the country, along with side bets on the cricket, with disastrous results' – are past. Under the ICC Code of Conduct, Mark Waugh will have to stick to the horses in 2001.

This Australian team also contains the usual solid core of golf enthusiasts: Ricky Ponting is an especially talented exponent, having carded 77 round St Andrews on Australia's last visit. This is a tradition spanning at least eighty years: Warwick Armstrong's men were guests at Gleneagles in 1921, the captain being characteristically unperturbed when his first drive travelled six inches. Many since have followed in his footsteps/divots. In 1930, Arthur Mailey penned a whimsical account of Australian cricketers at the links, where scorecards were scorned as looking 'too much like evidence' and games were judged 'according to the number of balls lost'. In more recent times, Allan Border described Merv Hughes as 'the worst male golfer I have ever seen'.

There is bound to be rapt attention on the British Open, which coincides with the Lord's Test. Should Australians be in contention on the final day, check for subtle signals. The 1989 team were so intent on the fortunes of countrymen Wayne Grady and Greg Norman that their twelfth man was delivering updates every second over during a tour match at Bristol, and only Tim Zoehrer was other than devastated when Mark Calcavecchia won the final play-off. 'I had £10 each way on him a fortnight ago,' he finally confessed. 'For the rest of us,' wrote Geoff Lawson, 'it was like losing a Test match.'

Steve Waugh's Australians were too late for the FA Cup Final, a staple entertainment of past XIs; on the way there in 1985, Allan Border had the experience of being serenaded with 'Tie Me Kangaroo Down' by a tube train full of Manchester United supporters. But they did have a chance to visit Wimbledon, following a precedent established in 1905 when Frank Laver and Charles McLeod saw the very first Australian finalist Norman Brookes lose to American Laurence Doherty. Donald Bradman himself attended four tournaments: he saw compatriot Jack Crawford win the mixed doubles in 1930 with American Elizabeth Ryan, while his team of 1948 made a point of polishing off Surrey extra quickly so they could watch countryman John Bromwich play.

The kinship that Australian cricketers feel for national representatives in other sports is noteworthy. When Mark Taylor's Australians celebrated overhauling England at Headingley on the last trip, for example, their chosen companions were three members of the Canberra Raiders rugby league team from Australia, in England for the World Club Challenge. Taylor recalled: 'To have fellow Aussies along – and especially respected fellow Aussie sportsmen – seemed to heighten the celebrations, making them even more special.' Such customs also reflect an aspect of England that has not changed. If Australians cannot recognise it now as a cradle of common culture, they acknowledge it as a centre for sport. So perhaps, even in an era where Anglo-Australian ties are so loose and ill-fitting, a subject of indifference in England and irritation in Australia, there remains a certain something about an Ashes tour: England is *locus classicus* of the games Australians love. Sport might no longer be a vehicle of imperial understanding between the countries, but it represents a solid and irreducible remnant of shared heritage.

Wisden Cricket Monthly August 2001

Extra Cover

Batting
The New Golden Age

GOLDEN AGES, WHETHER IN ART or literature or sport, tend to define themselves in retrospect, usually being held up against a disappointing present. So it is a surprise to find cricket, even after the entertainment extravaganzas seen in Australia this summer, revelling so freely in the idea of a Golden Age of Batting.

It is true that there is now considerable statistical evidence to support the idea that we are in a batting bull market. Average runs per wicket during 2003's forty-four Tests was 34: the highest level since 1989. Run rates are virtually unprecedented: since 2000, every team has been scoring faster than in the 1990s, Australia by the factor of a third (3.85 versus 2.95). Swelled by fourteen double-hundreds in 2003, individual batting averages are inflating like share prices during the dot com boom. The question is: are they doing so with the same connection to reality? Watching Australia's attack pursue a profitless line two feet wide of off stump as India broke 700 during the Sydney Test, as enjoyable as it was to those fatigued by Australian pre-eminence, it was hard to avoid the feeling that the Golden Age of Batting involved a regression to the Stone Age in every other department of the game.

Golden Age of Batting, of course, is essentially a tautology. Golden Ages are always about batting. As Neville Cardus wrote of the first, in *English Cricket* (1946), the Edwardian era was one in which 'the great batsman could absorb himself in the perfection of his own art in the face of an attack largely reduced to a static mechanism'. The measures of law-givers and the attention of innovators was geared towards batsmen's welfare, often transparently, from the interdict against throwers to the introduction of the six – not forgetting the introduction of marl, so significant in improving and regularising the quality of pitches. The sole attempt to redress the favour bat enjoyed over ball, the proposal in 1900–01 to liberalise the lbw law in order to punish those batsmen who'd become adept at using their pads as a defensive bulwark, failed anyway.

Likewise in our new Golden Age. Jam-packed international schedules are good for batsmen: in the event of failure, there is always another innings round the corner. But they condemn bowlers to a regime of increasing toil: the temptation is to 'bowl smart', within oneself, from a shortened run-up with a minimum of experimentation, risking nought where injuries are concerned – though they seem to come frequently enough. Batting, too, is not an especially complex activity. You can walk into a game without extensive preparation and in time rediscover your touch. Bowling is more complicated. Yet international bowlers are frequently expected to return from inactivity and injury with minimal first-class match exposure: witness Brett Lee's groping for rhythm and straining for speed in Melbourne and Sydney having appeared in only one Pura Cup match since last season.

Part of the problem, of course, is generational. With the retirement of a bevy of outstanding new ball bowlers, every country is being challenged by a lack of bowling depth. When Matthew Hayden made his Test debut ten years ago, he immediately had his hand broken by Allan Donald. Restored to the Test side two years later, he was humiliated by Curtly Ambrose and Courtney Walsh. After such an initiation, no wonder he feasts on the likes of Andy Blignaut and Sean Ervine. And great as were the presences of the Border-Gavaskar Trophy, its absences were no less significant. When the series began, the first choice attacks of each team would probably have pitted McGrath, Gillespie, Lee and Warne against Zaheer, Nehra, Agarkar and Harbhajan. Only Agarkar played without interruption; the rest were *hors de combat* some or all of the time. The absence of even a few key bowlers in a series upsets equilibrium, allowing batsmen to sweat out the top-liners in expectation of easier pickings later.

All this, in a sense, we know. Cricket's calendar, like Mark Twain's view of weather, is something everyone talks of but nobody does anything about. Does the hegemony of batting, however, reveal a deeper malaise in the game?

While we have been busy rejoicing in the splendour of stroke play, bowling seems to have been becoming less ambitious, more stereo-typed. Its rhetoric is accented to nagging and negating – 'the corridor', 'the channel' and 'getting it in the right areas' – rather than beating and bewildering. This is so even among Australians, the self-proclaimed missionaries of attacking cricket. Our bowler of the seventies was Dennis Lillee: athletic, intuitive, extreme, expressive. Our bowler of the nineties and the noughties has been Glenn McGrath: taut, trained, restrictive, repetitive. His autobiography *Pacemaker* (2000) provides a fascinating insight into the times. The second half of the book analy-ses a string of Test batsmen from all countries and their respective strengths and weaknesses. In each case, without perhaps realising it, McGrath recommends bowling ... prepare to be amazed ... 'a nice tight line on off stump' or some slight variation thereof. We've come a long way since the great DK.

An unconsciously revealing section of Ricky Ponting's recent *World Cup Diary* (2003), meanwhile, involves Australia's preparations for a game against New Zealand. Fleming? 'Bowl tight in the corridor just outside off stump.' McMillan? 'Tight, short of a length bowling in the corridor outside the off stump.' Astle? 'Good, tight, short of a length bowling in the corridor outside the off stump, and spin.' Styris? 'Good corridor and length bowling.' Cairns? 'Long-hops and flippers.' No, only kidding. 'Tight corridor bowling.' Harris? 'Straight and tight in the corridor outside off stump.' Vettori? It can't be true, can it? Yes it can. 'Good tight line and length in the corridor.' Doesn't anyone bowl in the vestibule anymore?

What has changed? One of the key differences between the eras is the frequency of one-day cricket. We underestimate now how difficult that earlier generation found its assimilation. Though the first official one-day match was an Australian occasion and the proliferation of one-day cricket was an Australian concept, the notion of cramping and curtailing cricket sat ill with Australian players. Lillee reflected his countrymen's ambivalence when he wrote in *One-Day Cricket* (1980):

'I know it sounds un-Australian, and I almost find the idea offensive, but in limited-overs cricket we must learn to think negatively.' Of Australia's performance in the 1979–80 World Series Cup, *Wisden* noted: 'Greg Chappell made it clear he disliked this defensive form of cricket. He attempted to win his matches without resorting to negative bowling or spreading his fielders round the boundary.' In a sense, the notorious underarm conclusion to the third World Series Cup final of February 1981 was as much a reflection of Chappell's contempt for one-day cricket as of the shortcomings of his sportsmanship, a sort of *reductio ad absurdum* of its principles.

Those days now seem long ago. Jeremiads about one-day cricket are a thing of the past – and rightly so. Much of the energy and ingenuity we appreciate in Test batting now is a result of the insinuation of attitudes from the limited-overs arena. But Test bowling now seems increasingly pervaded by one-day gospels of containment – check out the way modern slip fielders applaud a maiden in which four balls have passed unmolested to the keeper – and that is not so welcome a development. The principles of Test and one-day bowling respectively are far more different than those of Test and one-day batting: you actually spend a lot of time in one-day cricket trying to prevent the ball behaving as you'd like it to in a Test match, like swinging and turning. If Test cricket is becoming a batsmen's game, it may be because it is becoming more like one-day cricket, which always has been.

The other influence on the conduct of international cricket – often underestimated but nonetheless profound – is coaching. It is sometimes forgotten that until twenty years ago the team coach was something players arrived in. Analysis of one's game was simple and informal but personal. In the Buchananite age of biomechanics, video and computer-aided analysis and sports psychology, self-evaluation and self-direction are de-emphasised, and spontaneity and flexibility are less important than sticking to the agreed game plan. Bowling attacks under such a regime tend to be reduced, borrowing Cardus's expression, to 'a static mechanism'; recall the English coach who chided his bowlers for bowling too many 'wicket-taking deliveries'.

The effect of the philosophies of coaching can be experienced in an hour of television, when you soon obtain a sense of how ruled cricket has become by increasingly arcane statistics. The English scientist Lord Kelvin once observed that 'what gets measured gets made', and this in cricket has always been true: would bowlers have bothered trying to follow five accurate scoreless deliveries with a sixth had there not been the statistical recognition of the maiden? But what, one wonders, is the psychological impact of new performance indicators that disaggregate runs conceded from back foot shots, behind square leg, off effort balls, etc? And of the new video jiggery-pokery, from the Hawkeye that can scrutinise a bowler's line over the course of an over, or the pitch schematics that map the bowler's length so its regularity can be studied? Is there in modern bowling, one wonders, too, too much attention now on process and too little on outcome?

It may seem like an act of apostasy, or party poopery, to question the foundations of this halcyon era of batsmanship. But even during golden ages, Gore Vidal once said, complaints are heard about the general yellowness of things. And as yet, it might be wisest to restrict ourselves to two cheers. Bear in mind that if the opposite were true, and batting averages and scoring rates were in decline, we would not be toasting a Golden Age of Bowling: we'd be wondering what the hell was the matter with everyone.

Wisden Asia February 2004

Batting

Less Is More

ALL RUNS ARE EQUAL, BUT SOME are more equal than others. To demonstrate what I mean, let me take you back a decade to the first ball of an Ashes series in Brisbane.

Phil DeFreitas – a *faute de mieux* choice if ever there was, for the alternative would have been Martin McCague – jogged in and released a delivery with all the force of a meringue. Under the circumstances, it might have been allowed to pass unmolested. Instead, the ball was not glimpsed again until its recovery from the gutter at backward point: the first of twenty-five fours in a scintillating Michael Slater century. The exasperated exhalation of the collective English press corps verily rattled the windows of the media box.

Slater, who announced his retirement in June, left the stage to the sound of his own feet, and his career figures of 5312 runs at 42.8 in Tests and 14,912 at 40.8 in first-class cricket do not immediately proclaim him as a batsman from the front rank. But he belongs to that group of players whose runs were always worth a little more than equivalent sums by others, for the snap and the dash of their acquisition, or the force of their presence and personality. We live in quantitative times, when Cricket Analyst informs you how many runs Sachin Tendulkar has scored behind point on days of the week with an R in them, and when Stats Guru can slice and dice a record in the blink of an eye and will probably shortly offer filters that compare 'Size of Endorsements' and 'Number of Groupies'. Yet the omniscience they offer can be misleading. The scorebook confides that Michael Slater hit the first ball of the 1994–95 Ashes series for four; it relates nothing of how eyes rolled, shoulders sagged and hearts sank among English players, spectators and journalists.

Slater's bravura 176 that day took 244 balls. He kissed his helmet's badge after reaching a hundred; he might well have kissed the ball to atone for dismissing it so peremptorily. It was a profoundly under-rated innings, for he made batting appear far simpler than it was, and proved for others: between his dismissal and a Hick–Thorpe partner-ship beginning late on the fourth day, wickets cost barely 17 runs each. But such was his nature. Seldom has cricket seemed on planes more separate than at the SCG at the start of 1999 when, as Australia strove to set England a strenuous fourth innings chase, Slater made a price-less 123 from 189 deliveries as ten team-mates eked out 67 from another 200. It was like watching a kid skateboard down a precipitous glacier while mountaineers slipped soundlessly down crevasses.

Slater's swashbuckling sometimes put Steve Waugh in mind of William Wallace, Mel Gibson's *Braveheart*. He might also have been known as Sleeveheart, for there was never any mistaking his mood. On the first day of the 2001 Ashes series at Edgbaston, for instance, England's 294 owed everything to a 103-run last-wicket stand which left them, momentarily, in good heart. Flailing like a drunk in a saloon, Slater quelled the insurrection, reaching 76 in as many balls by the close: it was a devastatingly effective psychological assault.

Had he but known it, Slater was playing his last Tests; he would not last the series. As his muscles were tautened and his torso thickened by gym work, his stroke play seemed to lose its touch and timing. His life, too, seemed to take an indulgent turn with the disintegration of his eight-year marriage and the affectation of a Superman tattoo on his shoulder, homage to rock idol Jon Bon Jovi. Mind you, Slater could in all likelihood have enjoyed every aspect of the Jon Bon Jovi lifestyle providing he hadn't started playing cricket like him as well. 'Living on a Prayer', though a sure-fire karaoke hit, is an unfit soundtrack for batting.

Who else might fall into the same category as a batsman whose statistics fail to do him justice? One player whose Test record corre-sponds surprisingly to Slater's – 5345 runs at 42.42 from one more Test with the same number of centuries – is Ian Chappell. And it is true that, despite his undiminished stature as a captain, he would probably not be among those first picked in the top order of an imaginary all-time XI. Yet, again like Slater, his runs were always disproportionately valuable. Michael Holding once told me he thought Ian Chappell won Australia the 1975–76 series against the West Indies. This was puzzling, given Greg Chappell's 702 runs to Ian's 449, but Holding insisted: 'Other batsmen could get on top of you, but Ian would embarrass you'; Greg, among others, 'got the benefit'.

Perhaps no player found responsibility and rank so becoming. The best of Chappell was seen in the pivotal roles at number three, where he averaged 50.94, and as leader, when he averaged a neat 50. It might have been higher still had Chappell not been addicted to the hook, the stroke by which he lived and died. But just as Steve Waugh's refusal to hook evoked his self-denial, so Chappell's refusal not to became an emblem of his defiance. Chappell nominates as his most important innings his 56 at Lord's in June 1972. Before the series, he had complained of having 'a gutful of bouncers', and declared that in future he would hook them, but had then been pilloried for losing his wicket to the stroke twice in the opening Test at Old Trafford. 'I had a stack of letters telling me to stop hooking,' Chappell recalled, 'including one from my grandmother – and she wasn't the greatest expert on the game.' Once bitten, never shy, Chappell hooked five fours and a defiant six off John Snow into the Mound Stand: Greg, allowed to settle, and at one stage scoreless for almost an hour, took up the mantle in a century he thought his best.

Sometimes it is a batsman's personal presence that deepens and broadens the impact of their runs. For his own part, Chappell has always thought highly of Sri Lanka's new deputy minister of industry, tourism and investment promotion, whom he considered 'a leader of men'. Arjuna Ranatunga's Test record stands out primarily for its longevity: he passed 50 on forty-two occasions in ninety-three Tests but transited to only four hundreds and averaged 35. But as a presence, testy, teasing Ranatunga punched considerably above his not inconsiderable weight, drawing opponents into individual battles that gave space and scope to team-mates. Records will tell you, for example, that Hashan Tillekeratne is an 8-runs-per-innings better Test batsman;

there is no doubt at whom Australians, over the years, preferred to bowl.

On other occasions – instances that spectators welcome – it is the charm and ease of a batsman that enhances his influence. It's somewhat baffling to discover that Pakistan's Majid Khan and Asif Iqbal averaged less than 39 in their Test careers – inferior to players as essentially unmemorable as Shoaib Mohammed and Saeed Ahmed, and on par with all-rounders like Majid's cousin Imran and Mudassar Nazar. For in the 1970s, Majid and Asif personified Pakistan's cricket: brave, mercurial magnificos.

Australian cricket was then advancing like a tide. Yet I recall Majid casually creaming the first ball of a series from Dennis Lillee to the Adelaide Oval fence at backward point and challenging the great Australian to dislodge his distinctive chapeau, then later in the same Test Asif masterminding an 87-run last-wicket stand, of which Iqbal Qasim scored 4, with his supple skills of placement. Neither set great store by technique. Tony Lewis once described Majid batting in the nets at Glamorgan from a stationary position to demonstrate that footwork was not integral to good batsmanship, and becoming steadily convulsed with laughter as he proved his point; he walked on nicks to the keeper, including once in Melbourne when no-one but Rod Marsh raised other than a half-interested glove. Asif's hands sat irreverently apart on the handle and he would work straight deliveries square on both sides of the wicket, his long loose-fitting pullovers suiting a figure perpetually on the move. As batsmen so completely themselves, however, they could scarcely be hemmed in by orthodox means. At numbers one and six and respectively, they bookended Pakistan's top order with distinctive flourishes.

One need not, of course, be nominally a batsman to add weight to one's performances. To enjoy a magnified impact, runs should ideally occur either at a time when they are badly needed, or accrue in a manner of maximum irritation to the opposition: thus are tailend runs often of psychological worth beyond their numerical value. In the 1970s and 1980s, England obtained sometimes remarkable lower-order leverage from Alan Knott, Jack Russell and John Emburey, not least because their homespun methods made them so exasperating to bowl to. Keepers Knott and Russell made seven Test hundreds, often against the run of play; off-spinner Emburey, who began his career rightly rooted to the bottom of the order, made almost 1184 runs in his last thirty Tests at an average of 34.

It was precisely because they batted low in the order that Knott and Russell could get away with batting as they did. Many a spinner's line disintegrated in the face of Knott's impish sweep, often from feet outside off stump, while quite a few fast bowlers cursed Russell's pedantic certainty around off stump. And if his team-mates never seemed still at the crease, Emburey was a batting minimalist. His arms were stiff, his feet seldom budged, his back-lift was perfunctory, and his top hand was about as material to his method as his right ear. What he had was the mental toughness and physical strength to erect himself as a stumbling block time and again. It was in other respects, alas, that Emburey's game deteriorated; Martin Johnson's melancholy judgement towards the end of Emburey's career was: 'The only thing that stands in the way of his elevation to all-rounder status is his bowling.'

If all these subjective judgements seem to over-praise attacking or unorthodox stroke play, the player whose runs are now worth most, I think, is one who is making a great many of them in a fashion manifestly pleasing to purists. Several players down the years have challenged the Australians with devastating stroke play: Brian Lara, V.V.S. Laxman, Hansie Cronje, Saeed Anwar, Chris Cairns. Rahul Dravid is perhaps the first to confound them by classically orthodox means. He impresses because he comes so obviously to stay: when he bats in the morning, he is planning for the afternoon; when he resumes after tea, he is subtly envisioning the next day. He is a batsman, as it were, of the third hour, to borrow Basil d'Oliveira's wonderful description of his own approach to batting: 'The first hour, I give to the bowler; the second, I take him on; after that, I make him suffer.' India's top order with Dravid is better than India's top order without Dravid by a factor greater than the individual. By playing so obviously his own game and treating all comers alike, he emancipates others to do the same. Remove him, for instance, and Virender Sehwag cannot play with the same abandon nor Sachin Tendulkar with the same self-absorption. Where four from Michael Slater was an act of bursting exuberance, four from Dravid is the exercise of rational judgement – something, again, that no scorebook will tell you.

Wisden Asia July 2004

Fielding
Ground Force

THERE ARE STILL PEOPLE AROUND WHO THINK modern cricket is no good. Those silly pyjamas. That ugly sledging. Where are the great swing bowlers of yesteryear? Why does nobody play the late cut? Egad, they don't shoulder arms the way they used to.

One thing, however, stops the jeremiahs in their tracks: fielding. Yes, you'll hear them admit, it *is* rather good. Some of those catches are really rather remarkable. And, my word, they do hit the stumps quite a lot nowadays. Mind you, in my day … Then they're off again. Let 'em go. And have a think yourself.

Fielding *has* improved. It's arguable that it's never been better. Which should be a subject of rejoicing, given that it's the activity every cricketer, great or humble, performs more of than anything else. Yet it's still the facet of cricket that devotees ponder least, and the last to enter a tragic's mind when he contemplates the state of the game.

Especially where one-day cricket is concerned – of which another bout is imminent, the CUB Series kicking off on 11 January – this makes no sense. In the compressed environment of limited-overs cricket, almost everything is restricted. Batsmen bat a maximum fifty overs, bowlers bowl a maximum ten overs, captains' dispositions are governed by restrictive circles. Yet those who lament the foregoing forget the one cricket skill that is unrestricted: no rule requires you to catch the ball one-handed, or to throw only when batsmen are taking a second run. Its value, indeed, is enhanced: batsmen run more gauntlets, matches can turn on mere misfields. Reductive of everything else, one-day cricket expands fielding: it flowers in the environment, like an orchid in a hothouse.

*

There is a simple reason why fielding is considered an ancillary cricket skill: in a game obsessed with statistics, it is not and cannot be effectively measured. Yes, there are lists of players with the most catches, but these are mostly meaningless, equivalent to run and wicket tables without averages. They measure success but not failure: no-one tabulates catches missed, or runs conceded through fumbles, or runs prevented by an agile save. You'll hear it said that Australia's Ricky Ponting, South Africa's Jonty Rhodes or New Zealand's Chris Harris is 'worth twenty runs in the field', but this is a guess. It relates nothing of the runs batsmen don't take out of apprehension, or those that aren't made because of a trapeze-artist catch that makes the difference between Sachin Tendulkar scoring 50 and 150.

Because fielding exists at this imponderable level, however, it has an almost metaphysical existence. Great fielding is directed towards collective rather than individual goals. Michael Bevan surprised an Australian team meeting a couple of years ago with an important insight: precisely because fielding is an activity performed for the benefit of others, it is a measure of a team's unity of purpose. Watch an Australian fielding effort, in fact, and it is that mutuality that shines forth, whether it's a sweeper diving desperately to save his bowler from conceding a boundary or an infielder darting to the stumps to prevent overthrows.

Australians have actually always been noted for the quality of their fielding. 'We are constantly having quoted to us the case of the earlier Australian XIs to this country,' wrote the Englishman Percy Standing in *Cricket of Today*, 'with their machine-like quality of fielding the ball and returning it to the wicket in one movement.' And that was a hundred years ago.

They have, moreover, always been noted for the hustling bustle of their approach. After Australia's first visit to South Africa in 1902, the home team's wicket-keeper Louis Tancred wrote: 'They demonstrated to the casual South Africans that nothing counted so much for success as strenuousness and intensity. In no instance did they relax their keenness; never did they permit themselves to lose sight of the principles of playing as hard as they could and giving no chance.' Words that could attach easily to Steve Waugh's modern ensemble.

Yet fielding was regarded in a similar light to the skills of batting and bowling. That is, some could do it, some could not. Those that could excelled. Those that couldn't were hidden, in boltholes like fine leg or mid-on. It is strange that perhaps the most important axiom about fielding was written 104 years ago in Prince Ranjitsinhji's *Jubilee Book of Cricket*: 'Fielding is the only branch of the game in which, if one tries hard, one can be sure of success.' Strange because, while there have always been outstanding individual fielders, the idea of a brilliant fielding 'side' was unheard of.

The first was Jack Cheetham's South Africans fifty years ago. Man for man, they were an ordinary team, and not expected in 1952–53 to tax an Australian team that had won twenty-four Tests and lost two since the end of the Second World War. So Cheetham decided that, if his men couldn't outbat and outbowl the likes of Hassett, Morris, Lindwall and Miller, they should at least surpass them in the field. He had South Africa's legendary rugby coach Danie Craven design a compulsory daily exercise routine, put his men through fielding drills of unparalleled intensity, and pored for hours over the equivalent of modern 'wagon wheels' in order to optimise his own field settings. The John Buchanan of his day.

The outcome was that a vastly superior Australian team was held to a two-all draw. The high point of the South Africans' endeavours came in the Second Test at the MCG. Keith Miller smashed a straight hit down the ground that looked a certain six. Russell Endean darted to his right on the boundary edge, sized up the approaching ball, then leapt feet in the air with his right arm fully extended to catch the ball in one hand. 'Good God!' Miller blurted. 'He's caught the bloody thing!'

Since then, countless captains and coaches have unconsciously imitated Cheetham. The simplest way to transform a modest team is by improving its catching, chasing, retrieving and throwing: recent examples include Bob Simpson with Australia in the 1980s, Dav Whatmore with Sri Lanka and Steve Rixon with New Zealand in the 1990s. It makes sense: if your bowlers tend only to produce half-chances, make sure they're taken; if your batsmen can only chase small totals, keep them small with agile saves and fast, flat returns. In the words of Ranji: 'A bad field is an eyesore to spectators and a millstone round the neck of his side. Out upon him for a nuisance to society!'

*

Some techniques of fielding we'll take for granted during the CUB Series were also originally South African. The one-handed pick up and throw in a single movement was brought to a perfect pitch by Colin Bland, a raw-boned Rhodesian farmer who picked out stumps from cover like William Tell on apple patrol. The uncanniness of his eye is evoked by a run-out he performed at Lord's in June 1965, when England's Jim Parks tried to save himself by the old trick of getting between the fielder and the stumps; Bland skimmed his return so flat and low that it passed *between the batsmen's legs* to catch him yards short. Had Bill Lawry been commentating, they would have needed the St John Ambulance to revive him.

Bill was actually there when the other South African brainstorm was first pioneered: the two-man boundary sortie, where one man chases and retrieves while a second throws. In the Fifth Test at Kingsmead in February 1967, Jackie du Preez flicked a Bob Simpson leg glance back from the gutter, Eddie Barlow threw, and Lawry was run out without facing a ball. Residents of Durban, spared a characteristic Lawry entrenchment, were ecstatic.

Much more about the fielding we now watch, however, is modern. Two interwoven factors have lifted its quality since the mid-1970s. The first is obvious: the rise of one-day cricket, which rewards virtuosity and punishes error. The second is, perhaps, less so: the rise of professionalism. When players stole time to practise and play from other jobs, they tended to stick to core skills. When they had time and scope to tackle all cricket's competencies, it was naturally the one hitherto neglected that showed the steepest improvement.

For this is what truly distinguishes fielding's past and present: not so much the rise of the fielding superstar, but the demise of the duffer. The general raising of the bar even incorporated such unlikely improvers as Geoff Boycott. On his first tour of Australia in 1965–66, Boycott was regarded as one of the poorest fielders in the game: clumsy, myopic and lethargic. On his last tour in 1979–80, he darted in from mid-off to execute a direct hit run-out. Not only that, but the dour Yorkshireman startled team-mates by running in to the celebratory huddle shouting: 'I did it! I did it!' Thirty years ago, a fast bowler like Glenn McGrath would probably have fielded predominantly with his feet, and seldom chased a ball in the same postcode as another fielder. Nowadays, he not only cannot, because a slight misfield could cost his team a whole tournament, but need not, for a sizeable ration of fielding will have been part of his practice routine since he tackled the game seriously.

With the chance to rehearse and to perfect came innovation. Fielding seems a natural activity, but now has techniques as evolved as any in batting and bowling: the boundary slide, the underarm flick, the relay throw, the bounce throw. Positions in the circle in one-day cricket have become as specialised as those in the catching cordon.

The great innovator over the past decade has been another South African, Jonty Rhodes, recently retired from Test cricket but an indispensable member of his country's one-day side. There have been better throws, and deader eyes, but probably never a superior stopper. Self-deprecatingly, Rhodes calls himself a 'glorified goalkeeper': glorified because, theoretically, it's simpler to stop volleys coming always from the same place.

Rhodes' technique is actually unorthodox. Where the standard manuals insist that a fielder should walk in with the bowler, Rhodes comes almost to a halt with both feet on the ground at the point of delivery, ready to dive either way. His practice takes the form of a batsman belting the ball at a fifteen-metre zone defined by witches hats that he tries to protect. A short man with a low centre of gravity, his great skill is quicksilver recovery: from a prone or semi-prone position he is as good a throw as most men on two feet.

Rhodes likes to quote Colin Bland, who once philosophised that good fielders were 'part of the attack'. A key ingredient of the South

African game plan during the last decade has, in fact, been to guarantee that Rhodes does more fielding than anyone: with Allan Donald and Shaun Pollock attacking off stump, he is stationed at point; when the burly Afrikaner offie Pat Symcox pegs away at leg stump, he patrols mid-wicket. If he doesn't add a batsman to his Jonty's Greatest Direct Hits collection, he frequently saves runs merely by his presence.

Of all Rhodes' many run-outs, one during a Test match at Durban in November 1992 stands out for historical significance. India's Sachin Tendulkar, marooned by partner Ravi Shastri, was hurled out by a throw like a sniper's bullet from cover point, short leg Andrew Hudson completing the execution. Except umpire Cyril Mitchley's finger did not rise; rather his hands described the shape of a square, seeking the advice of colleague Karl Liebenberg, studying replays of the game supplied by the South African Broadcasting Corporation. A light flashed thirty seconds later indicating that Tendulkar had become the distinguished inaugural victim of the third umpire. And while television cameramen may not have done Shane Warne and Scott Muller many favours, for the past eight years fieldsmen have had no better friends.

*

The third umpire has subtly revolutionised cricket. In days of yore, a fielder needed to find the batsman at least a foot or two short to be guaranteed justice. Now a centimetre is all it takes. That means, essentially, that a batsman challenging a fielder's arm and eye is running a further thirty centimetres than he used to. And, as your average fully laden slugger takes about four seconds to navigate a pitch, your average unladen stopper has an additional 0.06 seconds in which to ready, aim, fire. It doesn't sound much, but when Jonty Rhodes or Ricky Ponting are in the neighbourhood, it's an eternity.

It used to be said that a throw actually hitting the stumps was wasteful, because of the risk of overthrows: the ideal was a return alongside the stumps that left a team-mate to break them. No longer. Ponting's pre-match drill is as simple as thirty throws at a set of stumps, which he'll aim to hit more often that not; preferably while a few opposition batsmen are looking on, a sort of preliminary shot across the bows.

For, cricket being a psychological game, fielding has an element of intimidation to it as surely as a bouncer or a big hit. In Adam Gilchrist's words: 'Few things lift a fielding team like a direct-hit run-out or even just a good throw that hits the stumps. It is as if a message has gone out through the players: "We're hot today and we only have to work a bit harder and throw ourselves around a bit and things are going to happen."' New Zealand's balding dibbly dobbler Chris Harris has made something of a speciality of this task. His pipe-cleaner man appearance camouflages an elastic fielder and mercurial character. In a Dunedin Test match played in polar winds against Sri Lanka four years ago, coach Steve Rixon sent Harris on as twelfth man just before a drinks break with the entreaty: 'It's cold as hell. Go and do something special.' Harris promptly threw out Hashan Tillekeratne, intercepted the drinks cart and drove it off himself. 'Will that do?' he asked his coach.

Sometimes, though, as Gilchrist says, all it takes is an early statement of intent. There was a classic cameo in this regard during the Sydney Ashes Test of January 1999. When England began chasing Australia's 391, Steve Waugh made a regulation save in the gully, then although the batsmen had not moved, nonchalantly threw down the striker's stumps. A murmur passed through the crowd and, it seemed, a heart murmur through the touring batsmen: in their eventual 227 lasting 80 overs, the Englishmen scurried only 41 singles.

Such scenarios, in fact, are not unusual. Paradoxically, we may be entering another evolutionary phase in fielding, in which its improved quality actually reduces the number of run-outs. An unnoted feature of the last World Cup was the marked decline in the number of such casualties. In thirty-six matches during the 1996 World Cup on the subcontinent, fifty-six batsmen were caught short. In forty-two matches during the 1999 World Cup in England, only forty-nine succumbed (and fourteen of those victims were Pakistanis, who may have been trying to run one another out anyway). Bowlers, of course, found English conditions more commodious during the last World Cup.

But it was surprising that, given the fielding talent on display, the tournament's decisive fielding contribution was Herschelle Gibbs' greasy-palmed pratfall.

It may be that batsmen, like drivers decelerating on a rain-soaked road, are dipping their lids to the likes of Rhodes and Ponting, and to the video vigilance of the third umpire, by running scared; or, at least, less boldly than before. In which case, this generation of cricket fans may become the tragics of tomorrow, telling our kids of the way Ricky Ponting seemed to run batsmen out at will, and receiving in our turn their nods of sympathy and disbelief.

Inside Sport January 2001

Slow Bowling

Think

A FEW YEARS AGO I WAS CAPTAINING a third XI on a windswept suburban oval, although in truth I might as well have been spectating. The opposition's number three, slumming it well below his class, had biffed and bashed his way to 70 off 40 balls. As I approached them, teammates were edging away, feigning distraction, hinting at injuries. 'Bugger this,' I thought suddenly. 'Time to give my offies a go.'

Anyone who's played Aussie club cricket will understand this to be a fairly desperate speculation on my part. On the flat, hard, dry pitches down under, real off-spinners don't eat quiche – they bowl it. As it happens, I began with a fortunate maiden in which I had the slogger's left-handed partner caught at slip. Nemesis arrived, though, well before hubris. After two coy defensive strokes the next over, the aforementioned number three belted me with the breeze for three fours. The towering six that then finished the over left me with lots of time to contemplate my next bowling change as long-on chased it down the road.

Spin bowling is worth thinking on. It's strange that when we ponder courage in cricket, it is usually that involved in its most elemental contest: the plucky batsman faced by bumping pitch, blinding light and brute bowling force. For the test of a spin bowler's nerves when a batsman is on the attack is just as great. The art's great underarm pioneer William Clarke was also the first to note its challenges: 'At times it's enough to make you bite your thumbs to see your best balls pulled and sky-rocketed about – all luck – but you must console yourself with "Ah, that won't happen again."' In a professional era, livelihood can also be on the line: thus Phil Tufnell's droll remark that every time he looked down the pitch he saw someone trying to end his career.

Graham Winter calls spin in *The Psychology of Cricket* (1992) not merely 'the most subtle art in cricket' but also 'the most psychologically demanding'.

The demands arise because of slow bowling's ineradicable tension: it invites the aggression that sometimes overthrows it. A spinner under attack then finds himself cast as fall guy. He has few of the counter-measures available to faster bowlers; Dilip Doshi once described the process of retarding Ian Botham's scoring as like 'trying to put the brakes on a sports car going at 250 mph'. Nor will the spinner experience much in the way of fellow feeling: attacking batsmanship, especially big hitting, sets everyone in a jolly mood. Andy Caddick can boast of trying to 'bore' batsmen out. Ashley Giles would never get away with a similar sentiment: his policy of negation in India set all heaven in a rage. Hypocrisy? Maybe – maybe not. But no wonder the breed is apt to feel put upon. 'It is said that everyone loves a lover,' wrote 'Ranji' Hordern, Australia's original googly merchant. 'But who, unless it be a big hitter, really loves a slow bowler? If he gets five or six wickets, there is no credit, because that is what he is there for. But if he gets "pasted" the crowd howl at him, his captain is furious, and even his team-mates are inclined to shun him.'

The nature of slow bowling is implicitly recognised in the old chestnut that it attracts a disproportionate number of romantic or idiosyncratic personalities, from Arthur Mailey and Chuck Fleetwood-Smith to Percy Fender and Cec Parkin. And it is true that slow bowling has had its mercurial individuals, unconsciously echoing the response of Australia's Sid Emery when Monty Noble told him he would be a great bowler if he could control his googly: 'If I could control myself I'd be a great man.' The longest section of Patrick Murphy's admirable survey *The Spinner's Turn* (1982) is devoted to 'The Characters', and dwells with relish over the likes of 'Bomber' Wells, Sam Cook, Tony Lock and Eric Hollies. Alan Ross's couplet from 'Watching Benaud Bowl' is among the most evocative in cricket verse: 'Leg-spinners pose problems much like love/Requiring commitment, the taking of a chance.'

The trouble with this is that there are as many counterexamples.

Warwick Armstrong, Wilf Rhodes, Hedley Verity, Jim Laker, Hugh Tayfield, 'Fergie' Gupte and Lance Gibbs – none of these were loveable cranks or whimsical vagabonds. Even Bobby Peel, excessive in most aspects of his life, was a strangely joyless charge for his Yorkshire skipper Lord Hawke: 'When at his deadliest and congratulated afterwards one could detect no gleam of pleasure on his countenance.' Nor does the popular precept that leg-break bowlers are nature's gamblers gain much empirical verification from, say, Grimmett and O'Reilly. And does Richie Benaud look like a man who'd be at home on a baccarat table at Monte Carlo?

Spin isn't for dilettantes either. Many have been famous for their solitary practice routines, from Hugh Trumble trundling away for over after over at a feather, to Roley Jenkins who once bowled for fourteen hours in the nets at Oxford in addition to playing in a three-day game. When Muttiah Muralitharan joined Lancashire in 1999, recalled Ian Austin, he startled his team-mates with his compendious memory: 'He knew more about Lancashire's record than the Lancashire players themselves. We'd be sitting in the dressing room or in the bar in the evening at an away game, and he'd suddenly start talking about one of our games from years back. He'd know all the facts and figures and couldn't believe that the rest of us didn't remember every last dot and comma of the game he was talking about.'

It might be truer to say that a spinner has to be unusually in tune with cricket, and its slings and arrows of outrageous fortune. Pitches can be prepared to suit slow bowlers, as they can be prepared to suit pace, but the best conditions in which to bowl spin are those that evolve naturally, thanks to wear, weather and chance. Even then, nature is unpredictable. On the first day of the Sydney Test three years ago, for example, Stuart MacGill harnessed a favourable breeze and a prematurely crumbling pitch in taking 7/104. Fish, firearms and cooperage came to mind. But the breeze dropped, the pitch perversely hardened, and he conceded 88 in the second innings without taking a wicket.

The slow bowler has, as it were, to 'go with the game'. No-one understands better than a spinner – dependent as he is on competent fielding, sympathetic captaincy and accurate umpiring – that cricket is a cumulative effort. No-one depends more keenly on the luck of the inch – in the matter of edges, crease lines and boundary lines – working in their favour. No-one needs to believe more unstintingly – in the face of sometimes overwhelming evidence to the contrary – that the game is ultimately even-handed.

Nor does anyone appreciate better than a spinner how humiliating the game can be. A batsman taking on a fast bowler is cricket's most vivid *mano e mano* contest – even the loser is part of the drama, like the losing swashbuckler in a swordfight. A batsman carving up a slow bowler inevitably seems more like bullying, violence seeming to assail gentleness. The victim, as Vic Marks recalled in describing bowling to Mike Gatting in full flow at Bath almost twenty years ago, is reduced to helplessness: 'Whenever he elected to defend one of my off-spinners, it was not through necessity but because of some whim, as if saying, "I probably could have hit that ball for six but I'm not a complete slogger, so I've decided to block it, as I'll be hitting another to the boundary in the next five minutes."'

If you're good enough, you can be generous about punishment. Bishan Bedi was justly famed for applauding sixes from his bowling. Despite the indignities to which Sachin Tendulkar submitted him in India in 1997–98, Shane Warne was able to write in *My Autobiography* (2001): 'It was a pleasure to bowl to him.' Such responses, though, are as rare as the individuals. Few spinners cannot have uttered inwardly the anguished cry of Ted Barratt, who bowled slow for Surrey twelve decades ago with a round left arm. 'Don't hit her sir!' he is said to have pleaded as he saw a big hitter tensing to uncoil. '*I'll give you four!*'

Those years ago on that windswept ground, the possibility of reviving this line crossed my mind. I'm ashamed to admit I also contemplated using the captain's prerogative. 'My word I'm tired all of a sudden,' I mused between overs. 'Could be time for a spell.' If it was good enough for Illy …

I decided, however, that 2–1–18–1 was not what I'd come out for. I resolved to bowl the slowest ball I could – to give my opponent time to think, and not, as some comrades insisted later, me time to duck.

The batsman, fatally conscious he was in the 90s, was bowled playing soberly back as the ball kept low. I'd seldom backed my ability on a cricket field before; for frankly good reasons, I still rarely do now. But just for a moment I glimpsed what it takes to prosper as a spinner. That and, errr, talent.

The Wisden Cricketer December 2003

Fast Bowling and Captaincy
Just Do It

IF YOU WERE TO PICK AS DECISIVE one of the many trends in cricket over the last century, it would probably be the rise and rise of fast bowling. A hundred years ago, the new ball was simply a piece of equipment necessary to the conduct of a match; today it is a team's cutting edge, lavished with care, attention, sweat and spit for the three or four pace bowlers in every attack.

Their emancipation, nonetheless, is incomplete. Fast bowlers might lead their attacks, but they seldom if ever lead their teams – it was, indeed, ever thus. Australia has given them especially short shrift. Ricky Ponting, Australia's forty-first captain, is the thirty-first batsman to hold the position. There have been four all-rounders (George Giffen, Monty Noble, Warwick Armstrong and Richie Benaud), three keepers (Jack Blackham, Barry Jarman and Adam Gilchrist), two finger spinners (Hugh Trumble and Ian Johnson) but only one fast bowler: Ray Lindwall, who was grudged a Test match when Johnson was ill, at Bombay in 1956. Never mind the ACB; such market dominance seems like a case for the ACCC.

Australia isn't the only country to have this taboo. In England, Mike Brearley believes, the discrimination is a vestige of the game's old class distinctions that vested authority in gentlemen, and kept horny-handed sons of toil to do the hard physical work. He recalls becoming aware of this in a school game against a Combined Services XI, which was composed of officers except for the opening bowlers 'Stoker Healey and Private Stead'. But other countries have at least hazarded the experiment: England (Bob Willis), West Indies (Courtney Walsh), India (Kapil Dev), South Africa (Shaun Pollock) and Pakistan (Imran Khan, Wasim Akram and Waqar Younis). And if

Australia is truly the country where Jack is as good as his master, why does it not seem to be so if Jack happens to have an aptitude for bowling quick?

The traditional theory has been promulgated by, among others, Sir Donald Bradman: that bowlers in general are incapable of the detachment necessary in determining when and how much to bowl. 'The difficulty with a bowler,' he wrote, 'is the constant fear that he will bowl himself too much and thereby incur an undercurrent of dissatisfaction amongst his colleagues, or underbowl himself because of undue modesty.' Brearley added that this was especially so with fast bowlers because their art is so strenuous and consuming: 'It takes an exceptional character to know when to bowl, to keep bowling with all his energy screwed up into a ball of aggression, and to be sensitive to the needs of the team, both tactically and psychologically.'

There is some weight to these propositions. To bowl or not bowl: that is certainly a question. And because fast bowling is physically enervating, the husbanding of energy it requires can breed a certain taciturnity. Even Ray Lindwall, wrote Ray Robinson, would grow tight-lipped while he thought through his bowling lures and traps: 'So intent was his watch that fellow players talking to him on the field got only monosyllabic answers, if that.'

The sort of scenario Bradman envisaged, nonetheless, would really apply only to a bowler about whom a team already nursed doubts, and there is no reason to think that batting captains are any less prone to selfishness or self-abnegation. One could argue the contrary, that a bowler-captain, accustomed to working batsmen out, and ideally placed to assess both his own physical state and the suitability of conditions to his methods, would be likelier to deploy himself at intervals most conducive to his probable success (nobody complains when batsmen captains do this, as when Steve Waugh continued to occupy his preferred slots at numbers five and six despite uncertainties higher in the order). And if a necessary attribute of leadership is the capacity to empathise with one's charges, who understands cricket's rigours more intimately than the individual whose job is the hardest yakka of all?

Essentially, it is an issue of image. 'The popular theory ... is that fast bowlers aren't all that bright,' complained Geoff Lawson. 'After all, how smart must you be to run thirty metres, 120 times a day, in the searing Australian heat?' Fast bowling is understood to be visceral, intuitive, spontaneous. Its definitive explanation is that *mot* of Jeff Thomson's: 'I just shuffle up and go wang.' The ideal captain is, or at least is imagined to be, rational and reasoned, thinking ahead rather than living in the instant. Here the most famous line is Richie Benaud's: 'Captaincy is ninety per cent luck and ten per cent skill ... but for heaven's sake don't try it without that little ten per cent.'

Fair? Scarcely. And the cool clinician is really only one type of leader. Another prerequisite of captaincy, Benaud once contended, is 'tremendous confidence'. In *On Reflection* (1983), he wrote: 'I believe that the man in charge of a cricket team ... must have the conviction that he can, at the very least, do anything he sets out to do; on top of this, he needs to believe that occasionally he can achieve the impossible.' And no-one embodied this more fully than a fast bowler, Keith Miller, whom Benaud classed 'the very best captain I ever played under'.

Miller was also a virile attacking batsman, of course, but his whole game was governed by the fast-bowling humours cited earlier: he was aggressive, dynamic, impulsive. And far from unsettling the New South Wales side that he led from 1952 to 1956, it galvanised them – and him. He made a century in his first match as skipper, took 34 wickets at 20 in his first season, then led them to three consecutive Sheffield Shields. His hunches kept his men guessing, and his opponents confounded.

Benaud describes how on the first morning of a game at the Gabba in November 1953, Miller summoned him to bowl his leggies in the sixth over. 'Nugget,' Benaud remonstrated, 'the ball's still new.'

'Don't worry about that,' countered Miller. 'It'll soon be old. Just think about the field you want ...' Looking at his young slow bowler's expression, Miller reassured him: 'It's all right, it'll spin like a top for an hour.' By lunch, Benaud had 5/17.

Like all distinguished captains, too, Miller was prepared to countenance defeat in the pursuit of victory. In February 1955, he set MCC an inviting 315 to win their tour match with more than a day to make them: a gauntlet that they were obliged to pick up and, at 3/145 with Hutton and May in command, with which they looked set to slap his face. Then having breezed in to bowl May, Miller set his spinners loose, who secured a 45-run victory. And for all its casual air and risky undercurrent, Miller's was a formula that proved remarkably robust: he was defeated as captain only thrice in thirty-eight first-class matches.

As the economist Maynard Keynes once observed, however, unconventional success provokes more suspicion than orthodox failure. When the Australian Board of Control deliberated on Lindsay Hassett's successor in November 1954, Miller was passed over in favour of the conservative Victorian Ian Johnson. And while obscure, the reasons that Miller became what Benaud has called 'the best captain Australia never had' are not difficult to guess at. 'He's too wild,' the Tasmanian cricket potentate Harold Bushby once confided, and by prudish 1950s standards, when one could be controversial for crossing the road against the lights, he was too hot to handle. Then there was the attitude of the chairman of selectors at the time. It may have been with Miller in mind that Bradman, in the chapter concerning captaincy in *The Art of Cricket* (1958), made clear that 'sheer sporting prowess is not regarded as the sole qualification for a player' and that a captain's 'private life must be beyond reproach'.

Since Miller, most Australian states have dabbled in fast-bowling skippers, if only in emergencies, from Dennis Lillee (two wins, one draw, no defeats) and Jeff Thomson (three wins, five draws, five defeats) to Geoff Dymock (one win, seven draws, three defeats) and Carl Rackemann (one win, five draws, four defeats). Few have given the impression of natural gifts. Rodney Hogg observed drolly after his only game as captain in November 1980 that it could be a lonely job: when he sought team-mates' advice as India were making 9/456 in their first innings, he noticed them edging away; when India folded for 78 in their second innings, everyone had a suggestion.

Miller's reputation, funnily enough, may have done as much harm as good to the advancement of fast-bowling skippers. They are still perceived as the bold alternative when a choice from among the batsmen is not obvious, and if they do not succeed are construed as a reflection on the concept rather than on either the individual or the team. In fact, with the exception of Imran Khan, none of the bowling captains to which other countries turned during the 1980s and 1990s was notable for their leadership flair. But this no more invalidated the idea of the fast-bowling captain than Kim Hughes' record means that nobody from Western Australia should ever lead Australia again.

Miller's experience may be instructive in different ways. He was fortunate to be leading a supremely competent team: Benaud, Arthur Morris, Norm O'Neill, Bob Simpson et al. A more callow bunch might have been disarmed by his mercurial temperament and laconic cue for the field to disperse: 'Scatter.' His gift was pushing established talent beyond its existing limit by his innate aggression and inspirational example. His experiences found an interesting echo in those of Geoff Lawson, who led New South Wales into three Sheffield Shield finals between 1989 and 1992 with similar imagination and initiative.

Lawson was strongly convinced that bowlers should make naturally competent captains, analytical faculties honed by experience with the ball: 'I spent many hours down on the fine leg fence over the years and I used that time to play the game of captaincy constantly. It may have been a shock to many observers that I made an excellent fist of the job when I got it ... but really I had been preparing a long time.' Being a fast bowler, moreover, made him aggressive: 'It is the nature of bowling, and fast bowling in particular, that you have to be aggressive to be good.'

Lawson noted in *Henry* (1993), though, that fast-bowling captains required of their colleagues a minimum level of cricket sense, because their supervision, at least while the captain was bowling, was inevitably not total. 'When you're running in to bowl,' he wrote, 'your concentration should be focused on where you want the ball to go – and not whether your fine leg has swapped over for the left hander.' Lawson depended, especially in tight one-day matches with their arithmetic challenge to permutate allocations of overs, on 'the co-operation and concentration of all players all of the time'. With the likes of the Waughs, Mark Taylor, Mike Whitney and Greg Matthews in his team, he seldom had to do without it. 'They were fine allies, always forthright in their advice – and I appreciated it greatly.'

There are new forces, however, that militate against the rise of the fast bowler cum captain. One-day matches, now such an important form of cricket, are essentially a batsman's game: it is hard to impose yourself on a match in only ten allotted overs. And, ironically, the rise and rise of fast bowling has had the effect of making its practitioners incomparably valuable. They are to be petted and cosseted, spared when possible, rested when necessary. Anything that smacks of an additional burden and might compromise their effectiveness is to be rigidly eschewed; even Lawson and more recently Paul Reiffel had to wait until their Test careers were over before they were entrusted with leadership roles. Having been thought too obtuse to do the job of captain, they might now be felt to be too important.

Inside Edge Fast Men: A Celebration of Speed 2004

The Future
A New World Order?

IT WAS SAID OF MIKHAIL GORBACHEV that he had a flair for walking backwards into the future. Cricket shares that talent. Coping with the present has been hard enough, without worrying about what's been round the corner.

The period in which *Wisden Cricket Monthly* was founded was a rare counterexample.* Having been elaborately repackaged and modernised in Australia during the intrusion of Kerry Packer's World Series, cricket arrived in the 1980s a few months early, having been gifted a market for games in a day, in a night, in eye-catching colours and tri-cornered tournaments, with more pizzazz for television, greater rewards for players, and advertisements between overs to bankroll both.

Two dozen years later, cricket finds itself in another bout of serious forethought, led this time by the International Cricket Council. The council is finding that the earlier revolution wrought most of the more obvious changes, and its activities have so far involved more of the same: more cricket, more television, more marketing, more money. The next wave of reform will be more fundamental, redesigning cricket for a rising generation of sports consumers, and priming it for pastures new.

Some years ago, I attended an Australian Cricket Board function at which chairman Denis Rogers unfolded a vision of a globalised game. With great solemnity and ceremony, he announced that Australia, as part of its ICC remit, would be taking cricket to China. One imagined a Lord's war cabinet with a world map on the wall: 'OK Australia, you

---

* This article was commissioned for the last edition of *WCM*, founded in 1979, as a consideration of the period of its history before it merged with *The Cricketer*.

take China; India, you take the rest of Asia; England, get South
America happening. The rest of you, spread out. Meet you back here
in twenty.'

Extending cricket's sphere of influence was never going to be that
easy, for three interwoven reasons. Its initial spread was as an imperial game. Its success sprang from its capacity to serve both colonial
and nationalist ends, to be a means for the payment of homage and
for the expression of independence. In post-colonial times, it is
drained of that meaning: it becomes simply a game to win, drawing its prestige from money and marketing. Invoking the riots at
Lambing Flat would only get you so far in building a Sino-Australian
sporting rivalry.

This throws the stress back onto the game itself. And, let's be
honest: much as we all love it, cricket is a hard sell. 'No thanks,' said
the pretty girl that R.C. Robertson-Glasgow, with 'misplaced kind-
ness', once invited to a game. 'Nothing ever happens at cricket; it is
just all waiting.' Of course, it only seems to be – but she had a point.
Cricket takes a long time. It can look spectacular, but isn't designed
for spectacle. It can entertain, but isn't calibrated as an entertainment.
The main action occurs further from watchers than in any other
sport; the complexity and eccentricity of its tenets and techniques are
not welcoming; many of its dottier rituals seem superfluous.

Five-day cricket, regarded by those who know as the game's paramount variant, is a particularly fiendish form in which to interest the
uninitiated. A weak international football team can thwart strong
opposition by throwing everyone behind the ball, aiming to grit out
ninety minutes for a scoreless draw, and might even net an upset goal
against the run of play; a weak Test XI, with 1800 minutes of available
time, will always get thrashed. As, at present, they are.

Cricket, in other words, takes a bit of effort to assimilate: it's the
party you realise is great after an hour in the kitchen when you find
there's heaps of beer in the bath and you know the songs they're playing, having survived the shock of encountering three ex-girlfriends
on arrival. Many of the game's subtlest and most confounding
aspects, furthermore, are intrinsic to it. Change them for the sake of

broader appeal, and you endanger not merely the goodwill of the existing community but the very qualities that distinguish cricket from other games.

It might be more helpful to render meaningful what is already there. There's nothing like seeing others enjoying a game, however strange, to encourage you to join in. At the moment, there doesn't seem a lot of enjoyment going round. 'International cricket feels flat, undramatic, even dull,' complained Scyld Berry in *Wisden*'s pages a year ago. 'Everyone is playing too much. Australia's pre-eminence in the Test and one-day game has become predictable ... "Cricket goes in cycles" is an adage only a fool will cling to.'

Having shrugged off its amateur past, of course, cricket must bear a certain burden of professionalised tedium. It has spent its inheritance of great players who learned their cricket the old-fashioned way, rising through the established grades, and playing at state, county and provincial level before higher honours. The generation that succeeded them, streamed into youth teams and academies as well as the first-class game, have been raised with different expectations: knowing that cricket could be their living, they've never needed to live for cricket. If the game today seems more routine, perhaps that should not surprise us. Has anyone paid money to watch you work lately?

What can we say, then, about a quarter century of professional international cricket? The trade-off was a necessary and unavoidable one: cricket could not withstand the tide of sporting commercialism. Television and sponsors had re-priced all games, and the remuneration of cricketers could not stand still. But, for players, it was only a partial emancipation. The attitude of boards of control since Packer has been unconsciously obedient of Alfred Hitchcock's advice regarding actors: 'Pay them heaps and treat them like cattle.' And some heaps have been taller than others.

It was this climate of mistrust and cynicism that smoothed the path to malpractice. When Sir Paul Condon's anti-corruption unit reported to the ICC on match-fixing in May 2001, it noted that players were 'not sufficiently involved in the administration of the game and ownership of the problems'. While the ICC doesn't have a great record taking advice from others, one might have thought it could take its own. Administrators have fared badly believing in cricketers' worst instincts; it might be more fruitful appealing to their better natures.

One would be even more emphatic about this had 'professional' not become so pregnant with meaning. It suggests diligence, dedication, attention to detail, as in 'professional qualification', but it also implies contrivance, conspiracy and sleazy expedient, as in 'professional foul'. Cricket is witnessing both: we have what might be called 'professional appeals', displays of calculated intimidation and petulance bearing no relation to the matter for adjudication, even 'professional catches', like the one for which Sourav Ganguly remarkably escaped censure during the World Cup final. These displays aren't evidence of an overabundance of high spirits, or of being supremely tough and competitive: they're just cheating. To make the most of the dividends of professionalism, cricketers themselves must confront some of its less appetising manifestations. It will not only be beneficial for the game; it will make the case for their influence in it unassailable.

We're at a hinge moment in cricket's history – a tipping point, to use the expression beloved of marketers and military men alike. A new age beckons; the trappings of the old are slipping away. But while cricket shouldn't walk backwards into the future, the occasional glance over its shoulder might still be useful.

Wisden Cricket Monthly September 2003